ADVANCES IN BIOLOGY AND MEDICINE SERIES

TYPE III SECRETION CHAPERONES: A MOLECULAR TOOLKIT FOR ALL OCCASIONS

ADVANCES IN BIOLOGY AND MEDICINE SERIES

Series Editor - Tsisana Shartava, M.D.

Parasitology Research Trends
Oliver De Bruyn and Stephane Peeters
(Editors)
2010. ISBN: 978-1-60741-436-0

General Anesthesia Research Developments
Milo Hertzog and Zelig Kuhn
(Editors)
2010. ISBN: 978-1-60876-395-5

Venoms: Sources, Toxicity and Therapeutic Uses
Jonas Gjersoe and Simen Hundstad
(Editors)
2010. ISBN: 978-1-60876-448-8

Biomaterials Developments and Applications
Henri Bourg and Amaury Lisle
(Editors)
2010. ISBN: 978-1-60876-476-1

A Guide to Hemorrhoidal Disease
Pravin Jaiprakash Gupta
2010. ISBN: 978-1-60876-431-0

Advances in Medicine and Biology. Volume 1
Leon V. Berhardt
(Editor)
2010. ISBN: 978-1-60876-863-9

Advances in Medicine and Biology. Volume 2
Leon V. Berhardt
(Editor)
2010. ISBN: 978-1-60876- 288-0

Advances in Medicine and Biology. Volume 3
Leon V. Berhardt
(Editor)
2010. ISBN: 978-1-60876- 015-2

Type III Secretion Chaperones: A Molecular Toolkit for all Occasions
Matthew S. Francis
2010. ISBN: 978-1-60876-667-3

ADVANCES IN BIOLOGY AND MEDICINE SERIES

TYPE III SECRETION CHAPERONES: A MOLECULAR TOOLKIT FOR ALL OCCASIONS

MATTHEW S. FRANCIS

Nova Biomedical Books
New York

For permission to use material from this book please contact us:
Telephone 631-231-7269; Fax 631-231-8175
Web Site: http://www.novapublishers.com

Library of Congress Cataloging-in-Publication Data

Francis, Matthew S.
Type III secretion chaperones : a molecular toolkit for all occasions / Matthew S. Francis.
p. ; cm.
Includes bibliographical references and index.
ISBN 978-1-60876-667-3 (softcover)
1. Molecular chaperones. 2. Bacterial proteins. I. Title.
[DNLM: 1. Molecular Chaperones. 2. Bacterial Proteins. QU 55.6 F819t 2009]
QP552.M64F73 2009 572'.633--dc22

2009048937

Published by Nova Science Publishers, Inc. ✢ New York

Contents

Preface

Common to many bacteria is the ability to establish a symbiotic relationship or to evade innate immune responses of an animal, plant, fish or insect host. Most often this capacity is mediated by a type III secretion system (T3SS). The function of these complex molecular machines is likened to a syringe-needle injection device that is dedicated to the translocation of effector proteins directly into target eukaryotic cells. Each translocated effector tends to possess a distinct enzymatic activity that aids in subverting host cell signaling for the benefit of the bacterium. Their translocation requires another class of secreted protein – the translocator – which form pores in the target eukaryotic cell plasma membrane through which the effectors may transit to gain entry into the cell interior. Most often, each secreted substrate requires a dedicated small, non-secreted cytoplasmic chaperone for their efficient secretion. Unlike traditional molecular chaperones, these specialized type III chaperones do not assist in protein folding and are not energized by ATP. Controversy still surrounds their primary role; as bodyguards to prevent premature aggregation or as pilots to direct substrate secretion through the correct T3SS. The later is supported by recent evidence that these chaperones can dock directly to the cytoplasmic face of the T3S machinery, possibly serving as a recognition motif for substrate secretion. Added to this functional complexity is their important contribution to system regulation, which can ultimately confer temporal order to substrate secretion. Moreover, some chaperones display a bewildering propensity to interact with several additional T3S-associated proteins – the relevance of which remains uncertain. Structural data has now appeared for several important type III chaperones, either alone or in complex with their cognate substrate. This is proving a fillip in our attempts to understand the mercurial ways in which these versatile proteins operate in nature. It is hoped that this article will provide information on type III chaperone function, as well as highlighting key recent advances in the field. May it also be a testament to the value of continued intense effort in unravelling the mysteries of type III chaperone biology.

Chapter 1

Introduction

Many bacteria at some point live their life in contact with vertebrate or invertebrate hosts. These associations may often be mutually beneficial. The host may provide a supply of growth nutrients or a safe haven in which bacteria are protected from harmful external environments. On the other hand, bacteria may breakdown complex bio-molecules that provide the host with usable by-products. This is exemplified by the soil bacterium Rhizobia that forms root nodules on legume plants. In this symbiosis, the bacterium receives carbon made available by the plant, while the plant receives nitrogen fixed by the rhizobia [1]. On other occasions however, bacterial-host associations may be of a parasitic nature, whereby bacterial colonization of plants, animals, fish or insects can lead to the onset of disease that may even be lethal for the host [2] In this regard, the emergence of bacterial pathogens could be an accidental consequence of mutualism gone astray. After all, their ability to induce a fatal disease in the host might also mark the end-of-the-line for the infecting bacteria. It follows therefore that regardless of the mutualistic or parasitic outcome, bacteria can use common molecular mechanisms to establish a frame-work for cross-talk with a host [3-8]. Among the large group of Gram-negative bacteria, a frequently used mechanism for communication with the host cell is the type III secretion system (T3SS) [9-11]

Chapter 2

Type III Secretion Systems

Biogenesis of a Functional Nanomachine

Bacteria harbor several different protein secretion systems for the purpose of moving cargo from the cytoplasm into or beyond the bacterial envelope [12]. Some of these are general systems required to fulfill normal physiological needs. Others, such as the T3SS, are more specialized in their use for intimate and direct communication with eukaryotic cells, thereby helping to facilitate meaningful bacteria-host interactions [13]. T3SS's are complex nanomachines, being composed of ~25 proteins that in a step-wise manner assemble together into a hollow conduit spanning the bacterial envelope [14]. Several of these core protein constituents are first secreted into the bacterial envelope via the universal *sec*-dependent secretion pathway. Polymerization of an extracellular appendage, or 'needle', extending from the bacterial surface signals the final phase of the assembly process. Secretion of these needle constituents are dependent on a working T3SS, which is established only after correct system assembly in the bacterial envelope. While T3SSs are a feature of many Gram negative animal-, plant-, insect- and fish-interacting bacteria, there morphology can have some striking differences. The most obvious difference can reside in the surface needle appendage. In the plant pathogens for example, the needle takes on the appearance of a much larger pilus-like structure. A conceptual illustration of some distinct T3SSs is given in Figure 1. This includes the protein secretion system for the biogenesis of flagella, which will be discussed later in this review.

During in vitro culturing of bacteria in defined laboratory media, these assembled T3SS nanomachines are competent for secretion of a diverse array of protein substrates into the extracellular milieu. Even though this offers a convenient means to study the biomechanics of type III secretion (T3S), this form of secretion is probably an in vitro artifact. The true role of T3SS is seen in response to target cell contact [15, 16]. Close contact is assumed to first trigger the secretion of a family of translocator substrates that assemble into a translocon pore in the eukaryotic cell membrane. Assembly of this pore is assisted by a 'needle-cap' protein that locates at the needle tip [17-22]. This T3S needle – translocon pore structure is believed to establish an uninterrupted connection between the bacterial cytosol and the host cell interior. Another set of substrates – the anti-host effectors, are then thought to travel through this conduit on their way to the cell interior. Although this dogma is widely accepted in the research community, ironically very limited experimental evidence is available to directly support it. To illustrate this, protein secretion through the needle conduit and release out through the distal tip has seldom ever been visualized [23-25]. Thus, despite intense effort, the precise mechanism of T3SS-dependent translocation of anti-host effectors into an infected eukaryotic target cell remains illusive.

Secreted Substrate Cargo

As stated, in response to target cell contact or growth in defined laboratory media, two major protein classes are secreted and released from the bacteria; the 'middle' and 'late' substrates. Late substrates constitute the anti-host effectors that are directly injected into target host cells through pores formed in the eukaryotic cell plasma membrane by the middle substrates – termed the translocators (Figure 1). Not surprisingly, all translocator proteins tend to harbor hydrophobic moieties promoting their integration into lipid membranes. However, how this translates into forming a functional translocon pore that can specifically recognize effectors and facilitate their translocation into the cell interior is not well understood [17, 26, 27]. Perhaps one or both translocator proteins interact with components of the host plasma membrane, such as cholesterol, which might facilitate their integration into the membrane to form the translocon pore [28, 29]. Pore formation also seems to be an active process because it is susceptible to feedback inhibition by the enzymatic activity of translocated effectors once they have reached a critical mass inside the eukaryotic cell [30-32]. This suggests direct interactions occur between the translocators and the anti-host effectors; either to guide their onward passage into the target cell interior or to establish feedback inhibition of pore formation[33]

In contrast, a considerable amount of functional data is available concerning the late substrates – the translocated bacterial effectors [34, 35]. These anti-host molecules possess a modular domain structure demarcating areas responsible for secretion, translocation and effector activity. Any given T3SS can be responsible for the specific transport of several effectors, all with different enzymatic activities – numbers as high as 50 are routinely reported for some plant pathogenic bacteria [36] Moreover, these molecules are often unique having no homology with each other. They were also acquired by bacteria separately from their cognate T3SS. Acquisition was probably through convergent evolution, pathoadaptation or direct horizontal gene transfer, since their effector domains encode for enzymatic functions that mimic many of the signal transduction processes occurring inside the host cell [37]. The translocated effectors therefore alter the activities of host cell molecules, thereby disrupting normal host signal transduction pathways for the benefit of the infecting bacterium. Thus, through the use of T3S effectors, bacteria have acquired multiple ways to try and out-smart its host. Discussions of these are not in the scope of this title. The interested reader can learn much more by consulting one or more of the suggested reviews [34, 35, 38-43].

A miscellaneous group of 'early' secreted proteins, and therefore distinct from the translocator or effector families, are also targeted for T3S. Compared to the latter two substrate families discussed above, these are almost certainly the very first substrates secreted by a newly built system. However, several of them have undefined roles in the secretion process and are probably not translocated into the eukaryotic cell cytosol. As mentioned already in this chapter, the needle constituents are polymerized at the bacterial surface following their T3S [27]. Their secretion and polymerization into a needle structure must occur before most other substrates can be secreted. Therefore, these needle constituents represent a subset of 'early'secreted substrates (Figure 1). Other early type III secreted proteins fall into a heterogeneous group of regulatory anti-activators [44, 45]. Their secretion is triggered by inducing signals such as target cell contact. The signal sensor might even be a fully assembled needle complex [27]. Lowering of the anti-activator concentration in the bacterial cytoplasm through T3S serves to relieve system repression, which then stimulates elevated synthesis and secretion of middle and late T3SS substrates.

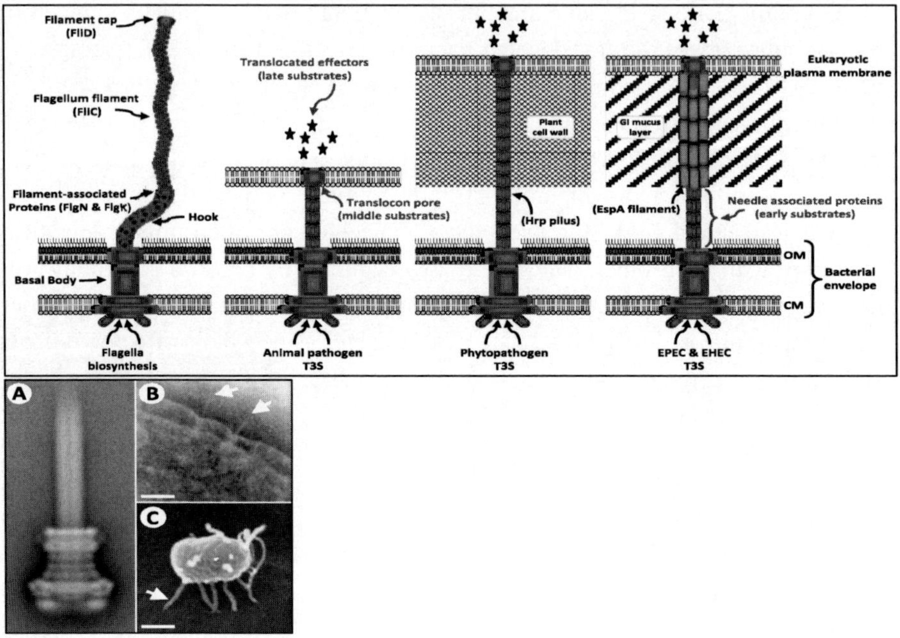

Figure 1. A schematic diagram of flagella and non-flagella T3SSs. The syringe-needle complex of non-flagella systems is reminiscent of the flagella hook-basal body structure. The typical non-flagella T3SS of an animal pathogen possess a short needle that connects with the eukaryotic host plasma membrane. Variants of this structure exist in plant interacting bacteria that possess an extended needle termed the Hrp-pilus for traversing the plant cell wall. In addition, pathogenic E. coli form an EspA filament on top of the needle that might be needed to penetrate through the mucous layer in the gastrointestinal tract (GI). Red highlight indicates the three substrate categories (early, middle and late) secreted by non-flagella T3SSs. This schematic was inspired from a review by He and colleagues 399. Panel A: An electron micrograph refined by difference mapping of negatively stained T3SS syringe-needle complexes purified from wild type S. flexneri. The image is part of a study performed by Boekema and colleagues 289 and is reused with permission "Copyright (2007) Elsevier Publishers". Panel B: An electron micrograph of negatively stained, osmotically shocked wild type S. enterica Typhimurium with the positions of the T3SS syringe-needle complexes in the bacterial envelope indicated by arrows. The image is part of a study by Galan and colleagues 400 and is reused with permission "Copyright (2000) National Academy of Sciences, U.S.A." Panel C: Scanning electron micrograph of enteropathogenic E. coli adhered to red blod cells via the EspA filament (arrow) associated with a T3SS. The image is part of a study by Knutton and colleagues 401 and is reused with permission "Copyright (2001) Wiley-Blackwell Publishing Ltd." OM – outer membrane, CM – cytoplasmic membrane.

This coordinates type III gene expression to various 'quality control' checkpoints along the path of appropriate T3SS biogenesis. Through necessity, this regulatory design restricts secretion substrate synthesis until a time when they are most needed, such as during the assembly process or as a translocation substrate for injection into eukaryotic cells [44, 45]. This 'anti-activator' concept will be revisted later in the article.

Chapter 3

Quality Control of Type III Secretion: On the Role of Chaperones

A Molecular Chaperone Perspective

The Cytoplasmic Heat Shock Response

To safeguard against the dangers of protein misfolding and aggregation, all organisms produce various families of molecular chaperones and proteases that participate in essential protein folding and degradation control [46, 47]. This inevitably ensures that proteins fold correctly and therefore maintain functionality. When encountering extreme physiological stress, the roles of chaperones and proteases assume even greater importance [48]. Their production is elevated significantly in a phenomenon known as the heat shock response. Briefly, the bacterial cytoplasmic response to stress is controlled by the σ[32] factor [48, 49.] The chaperones produced by bacteria grown in these conditions have variable size range (~10 kDa to ~100 kDa), but most function with the help of ATP hydrolysis to prevent aggregation and assist with protein folding or refolding in the bacterial cytoplasm. Apparently, unfolded or misfolded proteins expose hydrophobic regions that act as the catalyst for chaperone intervention [48]. Even though the majority of newly synthesized proteins tend to fold correctly with only minimal chaperone input, the remainder (perhaps fewer than 20%) are dependent on chaperone activity. Misfolded substrates not amenable to refolding are targeted by proteases for degradation [46, 47, 50]. Coincidently, the heat shock response is known to affect expression levels of T3SS genes encoded by various bacteria. This control of T3SS gene expression is often mediated by Lon- or ClpXP-dependent proteolysis of important regulators specifically involved in the control of T3SS synthesis [51-56].

The Extracytoplasmic Stress Response

Quality control of protein folding beyond the cytoplasm – in the bacterial envelope – can also require input from protein folding and degradation factors. This is especially needed when under assault from so-called extracytoplasmic stresses that can compromise bacterial envelope integrity and protein folding in the periplasm. Such periplasmic-located quality control factors function in an ATP-exclusion vacuum, which necessitates different molecular mechanisms to overcome issues of protein folding and refolding in the periplasm. Their production is under the control of two extracytoplasmic stress responsive pathways; the σE factor and the Cpx two-component phosphorelay system [48, 57]. Interestingly, T3SS gene expression and/or subsequent component assembly in the bacterial envelope also requires input from the σE and Cpx pathways [Liu and Francis, unpublished data] [58-65]. However, it is not yet clear whether this involvement involves a direct interaction of σE or the Cpx regulator with T3S gene promoters. An indirect association is also plausible through the production of protein folding and degradation factors that could function to ensure the correct assembly of T3SS components in the periplasm. On this point, several studies support a role for protein folding factors, such as the periplasmic disulphide oxidoreductase DsbA, in T3SS assembly in the bacterial envelope [Francis, unpublished data] [66-71]. In addition, alteration of the stoichiometry of bacterial outer membrane components, such as LPS, lipoproteins, or integral outer membrane proteins, can also have profound effects on T3S [72-77].

Sec-Dependent Secretion and the Phage Shock Response

Many bacterial proteins do not function in the cytoplasm, but must first be exported through the cytoplasmic membrane to take up residence in the bacterial envelope. On most occasions, protein precursors predestined for export across the cytoplasmic membrane must remain in an unfolded state. This export competence is achieved through the interaction of newly synthesized polypeptides with a distinctive molecular chaperone, termed SecB [78, 79]. Among the large family of molecular chaperones, SecB is unique in its ability to carry-out two specialized functions. It functions as a chaperone maintaining a large and heterogeneous group of proteins in an unstructured form necessary for export. It also functions as an 'export pilot' where the unstructured proteins are delivered to the membrane-located Sec-translocon for export of substrates through the cytoplasmic membrane. Substrate delivery occurs because each SecB-substrate complex possesses exquisite binding affinity for the SecA ATPase, a cytoplasmic protein that provides some of the energy needed for export of each substrate through the integral inner membrane Sec-translocon [80]. So important is this *sec*-dependent secretion system to bacteria that many of its components are essential. Defects in *sec*-dependent secretion across the cytoplasmic (inner) membrane also activate another extracytoplasmic stress response pathway called the phage shock response (PSR) [81]. The actual role of the PSR pathway is still not well defined, but it is considered important for maintenance of the protein motive force established across the cytoplasmic membrane [82, 83]. As briefly stated, a functional *sec*-dependent secretion system is a prerequistite for the early stages of T3SS component assembly in the bacterial inner and outer membranes. It therefore follows that an intact PSR is also necessary for optimal T3S [84].

Thus, with this brief overview one feature is for certain; T3SS synthesis and assembly is inextricably linked to several important physiological stress response mechanisms of Gram negative bacteria principally involving important functions of molecular chaperones. Clearly multiple layers of input are required to maintain quality control during the synthesis and assembly of such an inherently complex multi-component nanomachine.

Type III Chaperones – A Brief Introduction

The remainder of this article will detail a unique feature of T3SSs – their use of cytosolic chaperones in quality control. Since being first discovered in the laboratory of Guy Cornelis [85], T3S chaperones have been the focus of extensive research that has and enlightened our view of the T3S process [86-90]. In the early stages it was tempting to draw parallels with the molecular chaperones – particularly SecB of the *sec*-dependent secretion system. However, they have since proven to be structurally and functionally distinct molecules. T3S chaperones do not contribute to protein folding or refolding, nor do their functions depend on ATP binding or hydrolytic activity. They are small (less than 20 kDa) proteins with an acidic isoelectric point. Their function is to form highly specific transient substrate interactions to promote efficient type III-dependent substrate secretion and translocation. The specificity of each interaction is acute with T3S chaperones tending to target only one, or at most a few, cognate type III secreted substrates. In addition, their gene location is often near that of their target substrate(s), linking chaperone and substrate expression in the bacterial cytoplasm. For convenience, chaperones associated with target-cell contact inducible T3SSs (also referred to the non-flagella T3SSs) have been grouped into four general classes based on their structural similarities and the known functions of their cognate substrate(s) [88]. A fifth class represents chaperones of the flagella T3SS. Despite these commonalities, amino acid sequence similarity among individual chaperones within and between classes is usually quite low. A list of currently recognized T3S chaperones and their corresponding substrates is summarized in Table 1.

Distinct Classes of Type III Chaperones

Class I Chaperones

Small class I T3S chaperones (~10-15 kDa) target those anti-host effector substrates translocated into the host cell interior. They can be further classified based on their specificity for only one cognate effector substrate (class Ia) or several effector substrates (class Ib). During the past few years, tertiary structures have emerged for several of these molecules, either alone [91-97]or in complex with cognate substrate [98-101]. Overall, these chaperones have a common mixed α/β fold. They function as homodimers, where a localized portion of a singular bound effector wraps around the chaperone surface in a non-globular extended conformation [98-101]. This gives a chaperone:substrate stoichiometry of 2:1. By and large, class Ia and Ib structures are remarkably similar despite their lack of amino acid sequence conservation (Figure 2). However, structural resolution of at least two class Ib chaperones, Spa15 and CesT, revealed a distinctive dimerization interface that generates different orientations of the monomer pairs [91, 94]. This exposes additional potential binding surfaces, which could be one reason why the class Ib chaperones (but not class Ia) display broad specificity for binding multiple substrates.

Table 1. T3S chaperones and their cognate targets

Organism	Class	Chaperone	Crystal structure	Suggested function(s)	Known secretion substrate(s)	Other binding partner(s)	References
Yersinia spp. (plasmid)	Ia	SycE (YerA)	Yes (alone and in complex with YopE)	Masking MLD (Stabilizer)	YopE effector	YscM2(LcrQ), YscE	85, 95, 96, 101, 137, 139, 159, 177, 213, 349
	Ia	SycH	Yes (complexed with YscM2)	Hierarchal Secretion	YopH effector and LcrQ (YscM1 and YscM2) negative regulator		100, 137, 236, 238, 239, 349
	Ia	SycO	No	Masking MLD (Stabilizer) System regulation	YopO(YpkA) effector	YscM1(LcrQ)	173, 242
	Ia	SycT	Yes (alone)		YopT effector		92, 93, 350
	II	SycD (LcrH)	Yes (alone)	Stabilizer (Partitioning factor) System regulation	YopB and YopD translocators	YscY, YscM2, YscE and TyeA YscO inner membrane component	104, 108, 111, 113, 142, 143, 155, 158-161, 240, 349
	Ia^{3}	SycN and YscB	Yes (complexed with YopN)		YopN regulator		102, 103, 351-353
	V	YscE and YscG	Yes (complexed with YscF)	Anti-polymerization	YscF needle	YscE with YscG, TyeA, SycE and SycD	123, 354, 355
	nd	LcrG	No	System regulation	LcrV needle tip protein		124, 125
	nd	YscY	No		YscX	LcrH	158, 160, 356, 357

Table 1. (Continued)

Organism	Class	Chaperone	Crystal structure	Suggested function(s)	Known secretion substrate(s)	Other binding partner(s)	References
Yersinia spp. (chromosome)	Ia	SycP	No		YspP effector		186
	II	SycB	No	System regulation	YspB and YspC translocators	YsaE activator	166, 232
Pseudomonas aeruginosa	Ia	SpcS (Orf1)	No		ExoS and ExoT effectors		358
	Ia	SpcU	No		ExoU effector		359
	II	PcrH	No	Stabilizer (Partitioning factor)	PopB and PopD translocators		147, 154, 162, 360, 361
	Ia[3]	Pcr2 (PycN) and PscB	No		PopN		362
	V	PscE and PscG	Yes (complexed with PscF)	Anti-polymerization	Each other and the PscF needle	Each other	120, 122, 159
	nd	ExsC	No	System regulation	ExsE negative regulator	ExsC inhibitor	223-225
Shigella flexneri	Ia	IpgE	No		IpgD		363
	Ia	IpgA	No		IscB		364
	Ib	Spa15	Yes (alone)	System regulation	IpaA, IpgB1, IpgB2, OspC3, OspB effectors	OspD1 anti-activator	91, 168, 233, 365
	II	IpgC	No	Stabilizer (Partitioning factor) System regulation Hierarchal secretion	IpaB and IpaC translocators	MxiE activator	109, 145, 157, 229-231, 366, 367

Table 1. (Continued)

Organism	Class	Chaperone	Crystal structure	Suggested function(s)	Known secretion substrate(s)	Other binding partner(s)	References
Salmonella spp. (SPI-1)	Ia	SicP	Yes (complexed with SptP)	Secretion pilot	SptP effector	InvC ATPase	99, 169, 216
	Ia	SigE	Yes (alone)		SigD(SopB) effector		94, 141, 368
	Ib	InvB	Yes (complexed with SipA)	Secretion pilot	SipA(SspA), SopA(SipF), SopE, SopE2 effectors		98, 191, 369-373
	II	SicA	No	Stabilizer (Partitioning factor) System regulation Hierarchal secretion	SipB and SipC translocators	InvF activator InvI inner membrane component	112, 161, 163, 228
Salmonella spp. (SPI-2)	Ia	SscB	No		SseF effector		374
	IV[4]	SseA	No		SseB pilus, SseD translocator		153, 164, 165
EPEC and EHEC[1]	Ia	CesF	No		EspF effector		375
	Ib	CesT	Yes (alone)	Hierarchal secretion	Tir, Map, EspF, EspG, EspZ, NleA, NleG, NleH, NleH2 and NleI effectors	EscN ATPase	94, 138, 214, 215, 376-379
		CesD	No		EspB and EspD translocators		118
		CesD2	No		EspD translocator		117
	IV	CesAB	Yes (complexed with EspA)	Anti-polymerization	EspA pilus, EspB translocator		114, 115
	nd	CesA2	No		EspA pilus		116

Table 1. (Continued)

Organism	Class	Chaperone	Crystal structure	Suggested function(s)	Known secretion substrate(s)	Other binding partner(s)	References
Chlamydia spp.	V	CdsE and CdsG	No		CdsF needle	Each other	380
		Scc2 and Scc3	No		Putative CopB translocator		381
		LcrH-2	No		LcrE(CopN)		382
Aeromonas spp.	II	AcrH	No		AopB and AopD translocators		110
	V	AscE and AscG	Yes (AscE)	Anti-polymerization	AscF needle	Each other	121
Bordetella spp.	Ia	BtcA	No		BteA effector		383
Edwardsiella tarda	II	EscC	No		EseB and EseD translocators		384
	nd	EscA	No		EseC translocator		385
Vibrio parahaemolyticus (T3SS-1)	Ia	VecA	No		VepA effector		386
Erwinia amylovara	Ia	DspB/F	No		DspA/E		387, 388
Pseudomonas syringae[2]	Ia	ShcA	No		HopPsyA		389
	Ia	ShcB1	No		HopPsyB1		170, 390
	Ia	AvrF	No		AvrE		391
	Ia	ShcF/AvrPphF/ORF1	Yes (ORF1 alone)		HopF/AvrPphF/ORF2		97, 392
	Ia	ShcV	No		HopPtoV		170, 393
	Ia	ShcM	No		HopPtoM		170, 394
	Ia[5]	ShcS1	No		HopS1, HopO1-1 and HopO1-2		395, 396

Table 1. (Continued)

Organism	Class	Chaperone	Crystal structure	Suggested function(s)	Known secretion substrate(s)	Other binding partner(s)	References
	Ia[5]	ShcS2	No		HopS1, HopS2, HopO1-1 and HopO1-2		396
	Ia[5]	ShcO1	No		HopS1, HopS2, HopO1-1 and HopO1-2		395, 396
	nd	HrpG	No	System regulation	HrpV		397
Xanthomonas campestris	Ib	HpaB	No	System regulation	AvrBs1, AvrBs3, AvrBsT, XopC, XopF1, XopJ and HpaA effectors	HpaA regulator HpaC and HrcV structural proteins	134, 135, 398
E.coli / Salmonella (Flagella biosynthesis)	III	FlgN	No	Anti-polymerization System regulation	FlgK and FlgL hook-associated proteins	FliJ inner membrane component FliI ATPase	294, 295, 297, 298, 304, 305
	III	FliT	No	Anti-polymerization System regulation	FliD filament cap	FlhC regulator FliI ATPase	294, 295, 298, 303
	III	FliS	Yes (complexed with FliC)	Anti-polymerization	FliC filament		291-293, 296
	III	σ^{28}	No	System regulation	FlgM anti-σ^{28} factor		302

The stated chaperones are not necessarily all present in the one strain. Their distribution is often pathovar and/or strain specific.

EPEC, enteropathogenic *Escherichia coli*; EHEC, enterohemorrhagic *Escherichia coli*

Two co-chaperones that form a heterodimer, which resembles class Ia structural classification

Shares some characteristics with the class IV CesAB chaperone [115, 153]. Does not contain the tetratricopeptide repeats characteristic of the class II translocator chaperone family [104].

Contain secondary structure more reminiscent of class Ia [133]

nd not determined

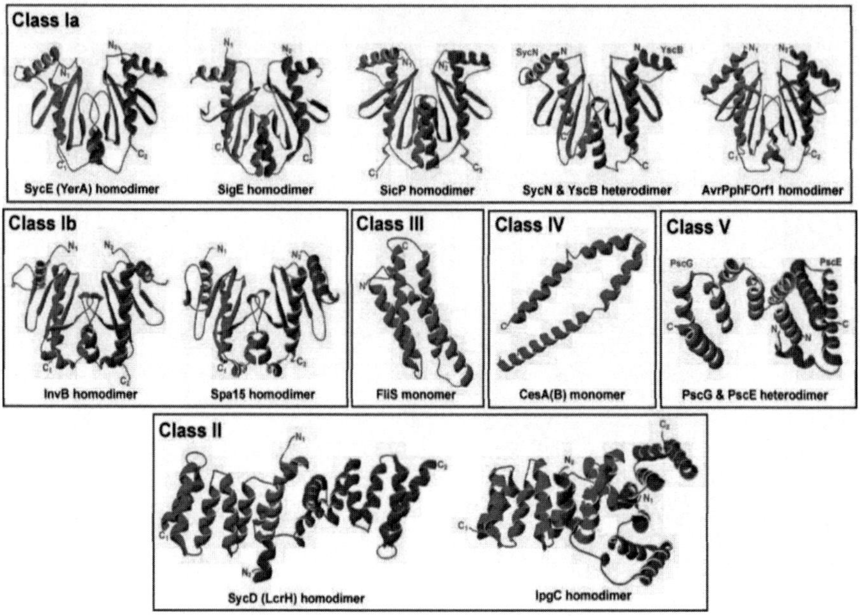

Figure 2. Ribbon models exemplifying structures of the various T3S chaperones associated with non-flagella and flagella mediated export through T3SSs. Models were drawn using the Swiss-PdbViewer 402 obtained from http://www.expasy.org/spdbv/. The monomers in the homodimer or heterodimer complexes are colored blue and red. Despite low sequence similarity, class Ia and class Ib T3S chaperones generally display remarkable structural conservation consisting of a tightly packed mix of β-sheets and α-helices. Significantly, these general structural characteristics remain conserved across diverse bacterial genera and are even maintained within the SycN-YscB heterodimer. However, other T3S chaperone classes are structurally distinct from each other and class I. For example, the analogous anti-polymerization function of class III and class IV T3S chaperones is not facilitated by similar structures and both function as a monomer; FliS adopts a compact four-helix bundle while CesA consists of a three helix hairpin. In addition, the class V T3S co-chaperone PscG that also plays an anti-polymerization role shares structural similarities with the class II T3S chaperone SycD possessing characteristic folds mediated by tetratricopeptide repeats. Class II T3S chaperones function as a homodimer although the two available crystal structures suggest the possibility of different orientations of the monomer pair – SycD is a head-to-head homodimer, while IpgC is an asymmetric homodimer. Examples shown are Yersinia pseudotuberculosis SycE (PDB identifier 1L2W; 139), Salmonella enterica Typhimurium SigE (1K3S; 94), S. enterica Typhimurium SicP (1JYO; 99), Y. pestis SycN and YscB (1XKP; 103), Pseudomonas syringae AvrPphF Orf1 (1S28; 97), S. enterica Typhimurium InvB (2FM8; 98, Shigella flexneri Spa15 (1RY9; 91), Aquifex aeolicus FliS (1ORJ; 296), enteropathogenic Escherichia coli CesA(B) (1XOU; 115), P. aeruginosa PscG and PscE (2UWJ; 122), Y. enterocolitica SycD (2VGX; 108) and S. flexneri IpgC (3GYZ; 109).

Additionally, a heterodimeric complex composed of the monomeric co-chaperones SycN and YscB function together as a novel class Ia chaperone for the regulatory protein YopN of *Yersinia* [102]. A 1:1:1 tertiary structure of the YopN-SycN-YscB complex revealed the SycN-YscB heterodimer to be structurally similar to the homodimeric chaperone conformation (Figure 2). However, the obvious asymmetry generated by the SycN-YscB association means that the configuration of YopN wrapped around the chaperone heterodimer is distinct from other class I T3S chaperone-effector complexes 103.

Class II Chaperones

The slightly larger class II chaperones (~15-20 kDa) bind the two pore-forming translocator proteins that are essential for the translocation process. A striking structural feature of this class of chaperones is their triad of tetratricopeptide repeats (TPRs) [104]. Each TPR consists of two anti-parallel α-helices in which alternating small and large residues enable the opposing helices to interlock to form a versatile scaffold for mediating various protein-protein interactions [105-107]. Structural models based on solved structures of TPR-motif containing proteins gave the first indication that class II chaperones adopt an all α-helical structure utterly distinct from class I chaperones [104]. The modeled structure also indicated potential concave and convex surfaces available for substrate binding. This has since been confirmed via experimentally derived structures of SycD (LcrH) from *Yersinia [*108] and IpgC from *Shigella* [109] (Figure 2). Not only did these structures confirm the twisted arrangement of the TPR motifs that gave rise to the modeled concave and convex binding surfaces, but they also confirmed a homodimeric conformation that had been suggested earlier [110-113]. Notably, class II chaperone dimerization is an essential feature, because a stable monomer unable to dimerize failed to rescue the chaperone defective bacterial mutants [108, 109]. However, it is not yet clear how this dimer is actually arranged given that the two solved structures are different – SycD existed as a head-to-head homodimer, while the IpgC homodimer was distinctly asymmetrical (Figure 2). Perhaps both homodimer forms are possible in nature and are biologyically relevant.

Miscellaneous Classes of Additional T3S Chaperones

The class IV and class V T3S chaperones (note that class III is reserved for chaperones of the flagella T3SS – see later), consists of a heterogeneous group of molecules that make up the remaining identified chaperones. At least conceptually, the class IV chaperone CesAB might be functionally related to other T3S chaperones, even though the only physical property it shares to other chaperones is a small size [114]. Nevertheless, CesAB binds in a 1:1 ratio with the EspA filament protein, a surface extension of the T3SS needle produced by enteropathogenic *Escherichia coli.* However, the structure of CesAB is entirely distinct from other T3S chaperones (Figure 2) [115]. Adding to this intrigue, EspA polymerization at the bacterial surface also requires another T3S chaperone, CesA2, which is a peripheral inner membrane protein [116] The extent of cooperation between CesAB and CesA2 is unknown. However, it looks comparable to the situation created by the highly aggregative translocator protein EspD that also requires two chaperones for secretion; the cytosolic CesD2 chaperone and the CesD chaperone, a peripheral inner membrane protein [117-119].

The PscE and PscG family of T3S co-chaperone heterodimers with specificity towards the T3S needle component, such as PscF from *Pseudomonas aeruginosa*, makeup the class V family [120]. Structural characterizations of members of this family revealed that all three proteins exist in a 1:1:1 ratio with the needle protein predominately in contact with PscG-like co-chaperone, while the function of the PscE-like co-chaperone is not obvious but it may stabilize PscG 121-123. Interestingly, the PscG family also harbor TPRs with folds similar to the class II chaperone family (Figure 2) [108, 122, 123].

Finally, in the *Yersinia* research field, some have also proposed LcrG to be a chaperone for the needle tip protein LcrV. This is based on the formation of soluble LcrG-LcrV complexes and the observation that LcrG improves LcrV secretion levels [124]. This is not unanimously supported however, given that LcrG still possesses important regulatory roles in the absence of LcrV 125. Nevertheless, this issue needs a resolution because homologues of both LcrG and LcrV exist in a sub-family of T3SSs encoded by a diverse group of important bacterial pathogens [10].

Analysis of various plant pathogens by whole genome and reporter-based screens has revealed many proteins with eukaryotic-like enzymatic motifs, all of which could be potential T3SS effector substrates [37, 42, 126]. Although a single strain may not encode every single identified effector, their repertoire can still reach over 50, which far exceeds the small effector substrate numbers associated with T3SSs belonging to pathogens of other hosts. Why plant pathogens need many more T3S effectors in comparison to other bacteria has been discussed in detail elsewhere. Those interested in the so-called 'gene-for-gene hypothesis' are encouraged to consult reviews on this topic, all which emphasize the fascinating and complex interplay between individual plant pathogens and their many susceptible and resistant plant hosts [37, 42, 127-132]. Probably, plant pathogens have developed extensive T3S chaperone networks to handle this vast amount of secreted cargo [10]. The occasional T3S chaperone has been discovered in plant pathogens that structurally resembles the class Ia variety; for example, avrPphF Orf1 from *Pseudomonas syringae* (Figure 2). However, some of these chaperones can differ from this usual paradigm in that they target multiple effectors that would normally classify them as class Ib T3S chaperones [97, 133] (Table 1). Another curiosity is the T3S chaperone, HpaB; to date the only known chaperone identified in *Xanthomonas campestris [*133]. Since HpaB supports the recruitment of several effectors to the secretion apparatus, it is considered to be a class Ib chaperone [134, 135]. However; in the absence of any other obvious T3S chaperone, an alternative possibility is that *X. campestris* uses HpaB as a global T3S chaperone for the export of all T3S substrates [135]. To develop this idea further, more must be learnt about HpaB biology. An obvious development would be solving the tertiary structure, either alone or in complex with various substrates, in order to make meaningful comparisons with other class I T3S chaperones.

Type III Chaperone-Substrate Recognition

Class I Chaperone - Substrate Interactions

Initial deletion mutagenesis experiments mapped the class I chaperone binding site to within the first ~100 amino acid residues of the cognate effector substrate [136, 137]. This has since been independently confirmed by various protease footprinting experiments whereby these chaperone binding domains (CBD) are protected from proteolytic digestion [98, 99, 103]. These findings are indicative of the CBD being one discrete modular domain at the N-terminus of effector proteins (Figure 3A).

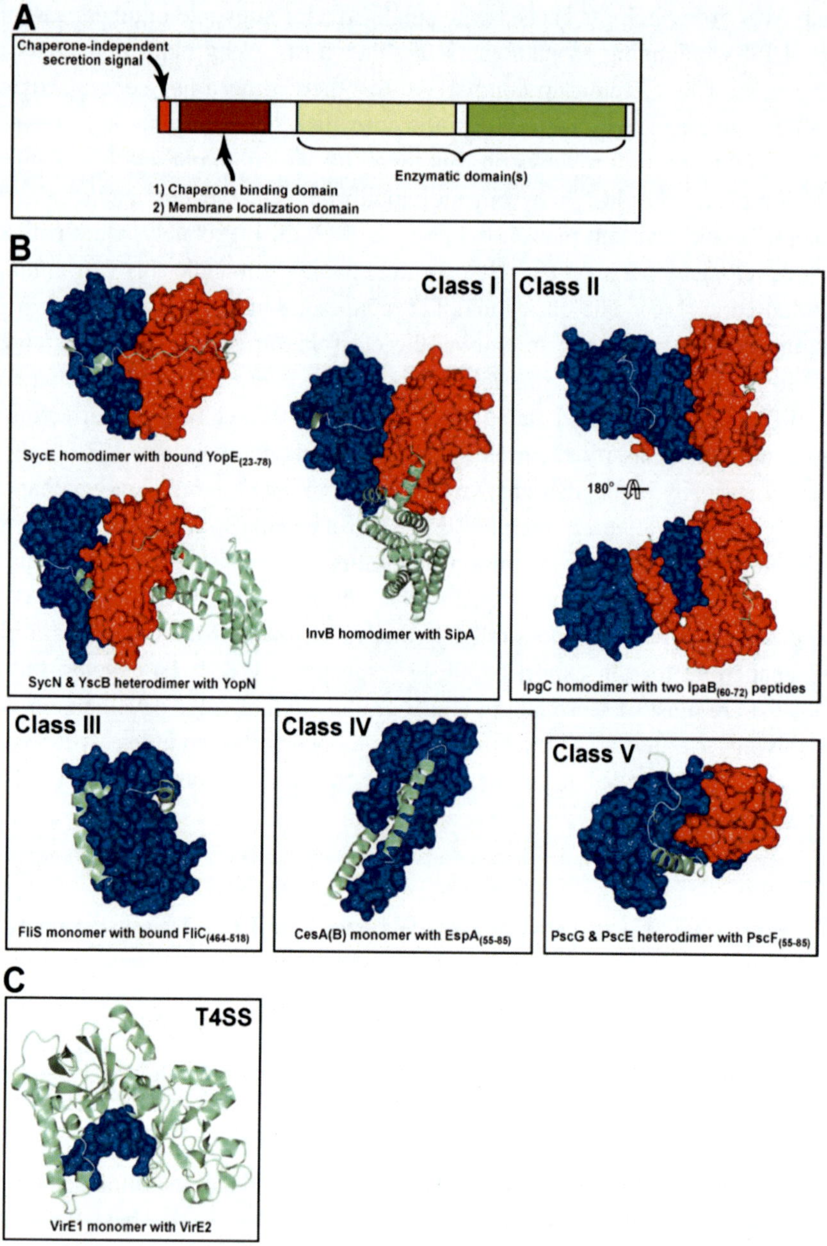

Figure 3. Structural characterization of T3S chaperone-substrate binding. (A) Translocated T3S effectors display a modular structure consisting of an N-terminal secretion signal, followed by the chaperone-binding domain (CBD) and one or more distinct effector

domains that possess enzymatic activity when translocated inside eukaryotic cells. The CBD also overlaps with a membrane localization domain in some effectors that is involved in targeting the protein to the correct intracellular location. (B) An effector substrate – class I T3S chaperone interacts with a 1:2 stoichiometry. The effector monomer uses its CBD to bind to their class I T3S chaperone homodimer. This induces a localized partial unfolding of the CBD region; even in the presence of bound chaperone, downstream effector domains remain fully folded and enzymatically active. Translocator substrates bind to class II T3S chaperones in a ratio of 1:1. Each monomeric unit of the chaperone homodimer engages the CBD of a separate translocator substrate. In the example given, the IpaB peptide is seen transcending the concave binding groove of the chaperone monomer. As with the effector substrate CBD, the bound translocator substrate CBD exists in an extended conformation lacking any teriary structure. While the structural context is quite distinict, a FliC filament also exists in a non-globular extended conformation when bound to the monomeric class III T3S chaperone FliS in a 1:1 configuration. In contrast, EspA associates with the class IV T3S chaperone CesA through extensive coiled coil interactions, but also in a stoichiometry of 1:1. Binding of the PscF needle to the PscG class V T3S co-chaperone also represents a distinct interaction interface; the C-terminal PscF α-helix is engulfed by the concave binding groove formed by the tetratricopeptide repeat fold of PscG. It also appears that the third member of the 1:1:1 trimeric complex, PscE, does not directly engage the needle filament. (C) The structure of the T4SS chaperone VirE1 is a simple singular α-helix that is encased by a bi-lobed monomer of the VirE2 substrate. To some extent, this conformation is reminiscent of the FliS-FliC complex. Additionally, the manner in which EspA polymerization is inhibited by CesA could also explain how VirE1 prevents VirE2 polymerization 320. Models were drawn using CCP4 MG (molecular graphics) software 403 obtained from http://www.ysbl.york.ac.uk/~ccp4mg/ and with some helpful tips from Tobias Hainzl (KBC, Umeå University). Molecular surfaces of individual chaperone monomers are colored blue and red. The bound cognate substrate is depicted as a light green ribbon structure. Examples shown are already described in the legend to Figure 2, with the exception of Agrobacterium tumefaciens VirE1 and VirE2 (PDB identifier 3BTP; 320) and S. flexneri IpgC and IpaB (3GY1; 109).

This has gained even further support from a study where CBD's were successfully exchanged between effector substrates with retention of function [138]. In a few cases, the CBD has been co-crystallized in a complex with the cognate class I T3S chaperone [98-100, 103, 139]. Within each of these structures, the CBD from a singular substrate wrapped around a chaperone dimer as an extended, non-globular polypeptide, interacting with hydrophobic patches on the chaperone surface (Figure 3B). Whether the bound CBD is actually unstructured has been recently questioned however – in the only structural study to use an entire substrate, the unfolded CBD of free substrate actually assumed an ordered state upon chaperone binding [140]. Hopefully these inconsistencies can be overcome with improved methods to co-purify and co-crystallize complexes of chaperone with full-length substrate. Despite this, close scrutiny of all known class I T3S chaperone-substrate co-crystal structures revealed a common binding motif. Within a large hydrophobic crevice spatially conserved in each chaperone, intermolecular contacts were made with each cognate substrate via the insertion of a few hydrophobic residues that comprise part of a conserved β-sheet motif [98]. Mutagenesis of these key hydrophobic residues within the β-sheet motif of the substrate confirmed the biological relevancy of this chaperone-substrate interface [98]. Independent mutagenesis of the hydrophobic residues that compose the large hydrophobic crevice of the chaperone also confirmed their importance to the interaction with substrate [141].

The structural conformation of class I T3S chaperone is constant. Even in the absence of substrate, all class I T3S chaperone structures highlight a homodimeric or heterodimeric state with similar structural folds (Figure 2) and a surface dominated by electronegative charges with sporadic patches of pronounced hydrophobicity (Figure 4A) [91-101, 103]. Through defined mutagenesis, the hydrophobic patches on the chaperone surface have proven to be critical for substrate binding [94, 141]. In contrast, electronegative residues do not necessarily affect substrate binding, so they might be more important for recognizing a component of the T3SS, such as the system ATPase (discussed later) [141]. These findings tend to suggest that subtle differences in the surface distribution of these hydrophobic patches are a determining factor in chaperone-substrate binding specificity.

Class II chaperone – substrate interactions

Mapping the class II T3S chaperone binding domains within the translocator substrates has proven to be more difficult. Each T3SS generally exhibits two primary translocator substrates. These do not display any sequence similarity, but do contain similar structural characteristics [26.] The first corresponds to the YopB family, which can consist of an N-terminal coiled-coil domain followed by two hydrophobic potentially membrane spanning domains. The second corresponds to the YopD family that possesses one (or sometimes two) hydrophobic transmembrane domains often followed by a C-terminal amphipathic α-helical domain. Arguably, interactions between the *Yersinia* LcrH (SycD in *Y. enterocolitica*) T3S chaperone and the YopB and YopD translocators have been studied in most detail.

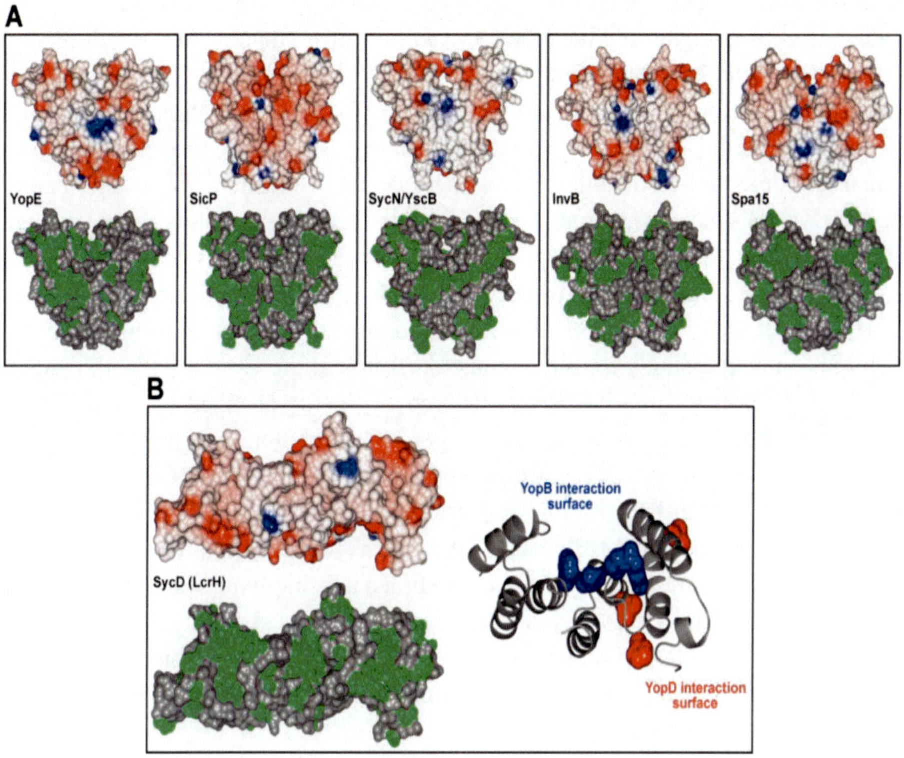

Figure 4. Determinants of the T3S chaperone-substrate interaction specificity. The molecular surfaces of representative class I (A) and class II (B) T3S chaperones have been colored according to their electrostatic (upper) and hydrophobic (lower) potential. An electronegative surface is in red, electropositive in blue and hydrophobic in green. Despite the structural orthology of T3S chaperones, each has a unique surface distribution of electronegative and hydrophobic patches. This key feature is considered to underpin chaperone-substrate specificity. Indeed, structure-based mutagenesis studies have confirmed the importance of hydrophobic and electrostatic residues in individual T3S chaperone function 154-156. The distribution of these hydrophobic and electrostatic residues in the class II T3S chaperone LcrH may also contribute to directing the binding of one substrate YopB, to the concave surface and the second substrate, YopD, to the convex surface of the tetratricopeptide fold 155. Models were drawn using CCP4 MG (molecular graphics) software 403 obtained from http://www.ysbl.york.ac.uk/~ccp4mg/ and with some helpful tips from Tobias Hainzl (KBC, Umeå University). Examples shown are already described in the legend to Figure 2.

While deletion mutagenesis analysis failed to uncover a singular discrete chaperone binding domain in YopB [142], a similar strategy suggested a large N-terminal domain (including the putative transmembrane domain) and the C-terminal **amphipathic** α-helix of YopD contributed to chaperone binding [113, 143]. Moreover, hydrophobic residues **within this amphipathic** α-helix were critical for the chaperone interaction [143]. The amphipathic domain is essential for YopD function, possibly contributing to YopD oligomerization [Costa *et al.*, unpublished data] [113, 144]. Perhaps binding by LcrH therefore prevents putative YopD self-association inside bacteria. From recent work, it is apparent that the AcrH chaperone from *Aeromonas hydrophila* also binds to its cognate substrates in a similar 1:1 stoichiometry [110]. The entire N-terminal half of AopB (a YopB homologue) is protected from limited protease digestion by bound AcrH chaperone. In addition, AopD (analogous to YopD) also uses the N-terminus encompassing a transmembrane domain and the C-terminal amphipathic α-helical domain to bind chaperone. Moreover, in a watershed structural study visualizing the *Shigella flexneri* IpgC chaperone interaction with IpaB (analogous to YopB), a N-terminal CBD motif incorporating the six residue PELKAP peptide sequence was identified in this substrate [109]. Actually, two CBDs were crystalised; one associated with each of the concave binding grooves found within the two chaperone monomers of the homodimer unit implying a stoichiometry of 2:2 (Figure 3B). Reportedly, similar peptide motifs also exist in the N-terminus of other translocator class proteins. This is consistent with IpgC binding to IpaC [145] and the *Salmonella enterica* Typhimurium SicA chaperone binding to SipB [146] only at their respective N-termini.

The translocator substrates are naturally aggregative, highly hydrophobic pore-forming transmembrane proteins capable of causing membrane damage if not controlled. For instance, over-expression of either YopB or the *Pseudomonas aeruginosa* PopB protein in the absence of their cognate chaperone results in growth cessation and eventual lysis of host bacteria [Costa *et al.*, unpublished data] [142, 147]. This nicely illustrates that an important role of the class II T3S chaperones is to conceal the hydrophobic transmembrane domains of their cognate substrates in order for them to be maintained in a stable and soluble form. It may also be relevant that some translocators display multiple functions – SipC and IpaC (YopD-family members) have both pore-forming activity in biological membranes and also exhibit intracellular effector functions [148-152]. In addition, a portion of YopD is also translocated into the host cell cytosol, which may imply an additional role(s) inside target host cells [144]. This suggests that the translocator proteins may possess discrete binding domains for interactions with protein partners of different origin and function. Perhaps an extensive coverage of the substrate by the chaperone is therefore needed to overlap with some of these possible binding sites. In this way, chaperone binding could minimize any premature intra- or intermolecular interactions from forming.

A few studies have investigated the molecular contributions made by the class II T3S chaperone in complex formation with substrate. The chaperone N-terminus has been suggested to contain information necessary for function, such as substrate binding or secretion [118, 153]. An *in silico* analysis of multiple sequences provided the first major breakthrough however, bringing about the discovery of a tandom array of TPRs in this class II chaperone family [104]. Using this structure as a scaffold for analysis of a large collection of mutagenesis data, our laboratory was able to confirm a role for these TPRs in chaperone stability, substrate binding and substrate secretion [104, 143, 154-156]. These data even revealed the possibility of simultaneous binding of chaperone to both substrates given the identification of two discrete substrate binding interfaces – YopB preferentially binds to the concave face while YopD attaches to the convex face of LcrH (Figure 4B). This possibility is supported by documented trimeric complexes of LcrH-YopB-YopD [142] and AcrH-AopB-AopD [110], although it appears unlikely to occur for IpgC, IpaB and IpaC [157]. Realistically, the only way to unravel these mysteries is to produce structural data of chaperone in complex with one or both substrates. This is a major technical challenge given the hydrophobic transmembrane properties of the translocators. However, a key advance in this direction has come with two recent studies. The crystal structure of LcrH/SycD was the first experimental structure of a class II chaperone specific for the translocators [108]. While this structure confirmed the presence of TPRs, it also demonstrated the functional relevance of SycD dimerization, formerly an underappreciated protein-protein interaction interface in T3S chaperone biology. The second study was able to co-cyrstalize the IpgC chaperone in complex with truncated IpaB (a YopB analogue) [109]. Not only did this study define the CBD within IpaB, but it also highlighted the extensive interactions that mediate chaperone substrate binding (Figure 3B). In particular, three substrate binding pockets were discovered in the concave binding groove on IpgC that are well conserved in class II T3S chaperones. These marry with the CBD contained within the IpaB translocator substrate. As such motifs are evident in other IpaB-related translocators, this is indicative of a conserved mechanism of substrate interaction for this class of T3S chaperones [109].

It is also notable that class II chaperones can possess other known interaction partners. For example, SycD/LcrH is also known to engage YscO, YscY, YscE and TyeA [111, 158-161], although the biological relevance of such interactions have not yet been sufficiently explored. Having the SycD crystal structure will now allow a directed exploration into these additional chaperone-substrate complexes as well. TPR arrays are highly versatile protein-protein interaction motifs able to contribute multiple binding sites for interactions [105-107]. Their conservation in class II T3S chaperones suggests that determining the full repertoire of chaperone binding partners and understanding the corresponding functional consequences of these interactions to T3S will be an area of fruitful investigation.

Other Chaperone – Substrate Interaction Classes

On the back of important structural studies, the molecular basis of binding between the class IV chaperone CesAB to the EspA pilus has been elaborated. Not only is the CesAB structure distinct, but the CesAB::EspA interaction interface is quite different from the chaperone-substrate interactions discussed above. It is dominated by coiled-coils contributed by both the T3S chaperone and the substrate (Figure 3B) [115]. Another distinct interaction occurs between the class V T3S co-chaperone PscG/YscG family and the needle component PscF/YscF. In this complex, the C terminus of the needle protein is buried within the concave hydrophobic groove formed by the TPR-containing co-chaperone (Figure 3B) [121-123]. The function of the third member of this trimeric complex, PscE/YscE, is intriguing for it clearly associates with PscG, but not the needle.

Chapter 4

Demarcating Type III Secretion Chaperone Function

One of the perplexing aspects of T3S chaperone biology is their apparent involvement in many T3S processes. Initially, T3S chaperones were broadly identified as bodyguards for their cognate substrates, ensuring their pre-secretory stabilization and efficient secretion. Such roles were initially likened to the function of the 'traditional' molecular chaperone SecB. During the last few years however, it has become apparent that T3S chaperones also ensure that their substrates are secreted through the correct T3S pathway, and perhaps even orchestrate a hierarchal secretion among the multiple substrates being secreted. Adding to this functional complexity is the clear evidence that some of these chaperones regulate gene expression and thereby create a molecular link between expression and secretion of T3S substrates. A summary of the major T3S chaperone functions are highlighted in Figure 5. Because most T3S chaperones tend to boast variations in their functional repertoires, there is some difficulty in pin-pointing their precise biological function. Such intrinsic functional diversity is probably a consequence of two issues; the unique physical and functional parameters displayed by the vast amounts of potential T3S substrates and the specific environmental niches where different bacteria employ their customized T3SSs. What follows below is an account of the most notable T3S chaperone functions. However, it is important to realize that these functions need not be mutually exclusive and that individual chaperones can exhibit multiple functions.

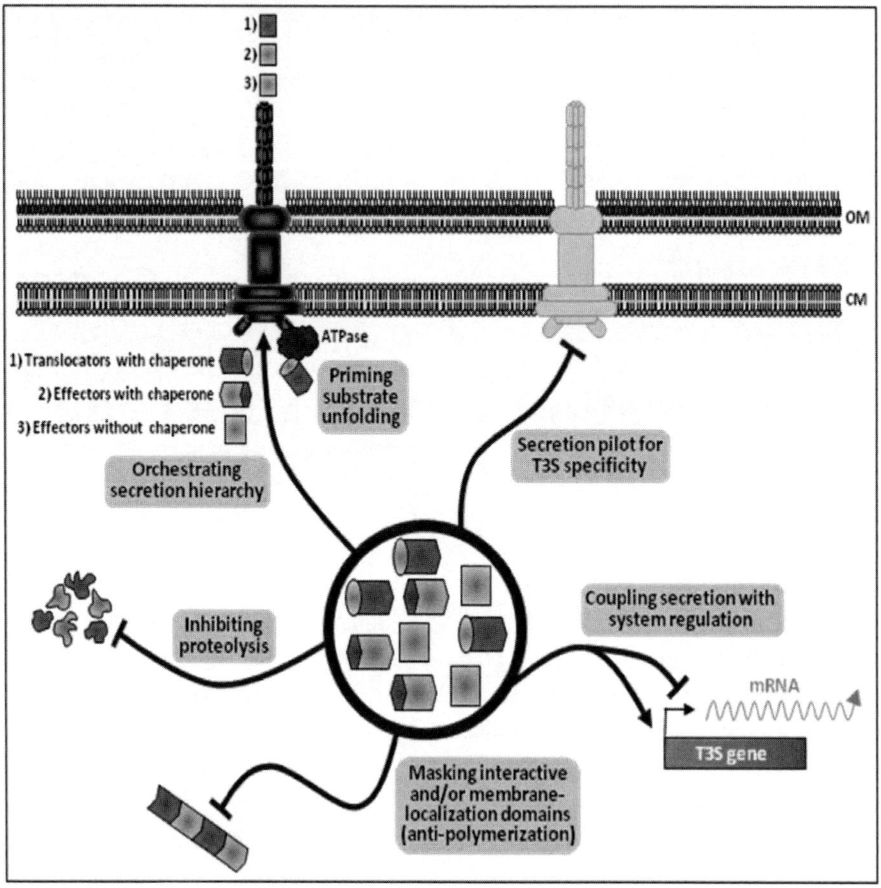

Figure 5. A schematic summary of the T3S chaperone molecular toolkit. At least six major functions have been ascribed to one or more T3S chaperones. Each function makes a significant contribution to efficient T3SS assembly and/or substrate secretion. The ratio of substrate bound to free chaperone is used to sense the status of T3SS activity that in turn influences the level of T3SS gene expression. Substrate bound by chaperone is also piloted to the cognate T3SS ensuring specific secretion through the correct secretion system. Bound chaperone can also be recognized by the system ATPase, aiding in localized substrate unfolding as a prerequisite for efficient substrate secretion. Specific recognition of different chaperone classes by a component of the T3SS probably also contributes to ordering substrate secretion to ensure that middle substrates (the translocators) are secreted before late substrates (the anti-host effectors). Finally, pre-made substrate pools are also stabilized in the bacterial cytoplasm through association with cognate chaperone. This prevents premature intra- or intermolecular interactions and confers resistance to endogenous proteolytic activity. OM – outer membrane, CM – cytoplasmic membrane.

Preventing Premature Interactions

Partioning Factors – Masking Substrate Interaction Domains

Arguably, one of the less contentious roles for T3S chaperones is that ascribed to the class II family – that of a partitioning factor to prevent premature association and subsequent degradation of the two translocator proteins prior to their secretion. This was first described in an elegant study by Menard and colleagues [157] whereby the IpgC chaperone stabilizes both IpaB and IpaC by preventing a premature IpaB-IpaC interaction in the cytoplasm of *Shigella flexneri*. Similar roles have subsequently been suggested for other chaperones of this translocator class [118, 142, 147, 153, 155, 162-166]. It remains to be seen however if this partitioning is achieved by simultaneous binding of both substrates to the same chaperone molecular unit. Based on our own studies with class II T3S chaperones, this seems conceivable since the PcrH and LcrH chaperones appear to contain two distinct substrate binding sites – possibly one for each substrate [154, 155] (Figure 4B). One also wonders whether the chaperone interacts with structurally similar domain(s) within each translocator protein and whether these domains are also responsible for establishing the translocator-translocator interaction. If so, class II T3S chaperones would function to partition the translocators by shielding these interaction domains. This possibility has been endorsed by a recent structural characterization of the IpaC N-terminus in which the IpaB and IpgC binding domains are believed to partially overlap [145]. It is also consistent with a conserved N-terminal chaperone binding motif, originally identified in IpaB, present in a variety of other functionally related translocator proteins [109].

While the role of partitioning factor is clear, it is also worth remembering that some translocators are toxic to bacteria when expressed in isolation [Costa *et al.*, unpublished data] [142, 147]. Thus, by binding to the substrate N-terminus, class II T3S chaperones also mask the hydrophobic transmembrane domains that elleviates potential substrate toxicity. This means that this T3S chaperone family is essential for maintaining pre-made pools of the pore-forming translocator proteins in an innocuous form inside the bacterial cytoplasm.

In contrast, only some class I T3S chaperones are needed to stabilize pre-made substrate pools [167-170] used for rapid deployment following target cell contact [168, 171]. While it is appreciated that pre-made substrate pools can have instant impact on the outcome of a bacteria-host cell interaction, it is not immediately obvious why some substrates are unstable and degraded in the absence of their cognate chaperone while others remain stable or do not even require a chaperone at any stage of their production or secretion. Quite probably, proteins involved in premature intra- or inter-molecular interactions in the bacterial cytoplasm are targeted for proteolytic degradation. Presumably, these premature interactions would normally be prevented by bound chaperone, as is the case for substrates bound by class II T3S chaperones (described above). It is therefore intriguing that non secreted pools of effector substrates are susceptible to Lon protease digestion in the *Pseudomonas syringae* phytopathogen lacking the corresponding cognate chaperones, whereas substrate-chaperone complexes are resistant to Lon-dependent degradation [170]. It is not yet clear why some T3S effectors are specifically targeted for Lon-dependent degradation. Intuitively, these data highlight a potential universal mechanism for pre-secretory stabilization of selected T3S substrates in all T3SSs; that of protection from Lon-dependent proteolysis via the action of T3S chaperones. As previously mentioned however, not all T3S substrates are prone to degradation since some are stable in the bacterial cytoplasm even in the absence of a T3S chaperone [172]. In these cases, the cognate chaperone is more than likely essential for substrate targeting to the secretory apparatus (discussed in the next section).

Shielding the Membrane-Localization Domain of Effector Substrates

It has also recently surfaced that several effectors possess a membrane localization domain (MLD) within this first 100 amino acids just downstream of the small chaperone-independent secretion signal. Significantly, the MLD overlaps with the CBD (Figure 3A) [173]. MLD's target translocated effectors to host cell membrane compartments in a location where they presumably engage their respective host molecular targets [173-176]. In the absence of a cognate chaperone, effector substrates harboring MLD's are prone to aggregate in the bacterial cytoplasm, while derivatives engineered to lack their MLD remain soluble even if they no longer bind to their cognate chaperone [173]. In fact, only in the presence of chaperone do MLD-containing proteins remain in a soluble form [139, 173, 177]. This work highlights another related function of some T3S chaperones; shielding their cognate substrate MLD's to avoid aggregation before they are successfully translocated to the eukaryotic cell interior.

Anti-Polymerization Factors

T3SSs are characterized by extracellular appendages that protrude from the outer membrane in much the same way as the bacterial flagella. In phytopathogens, this takes the form of a pilus-like structure, while enteropathogenic *E. coli* have a filament that extends out from the typical needle present on all other cell-contact inducible T3SSs (see Figure 1). These three types of T3S-associated structures consist of a highly aggregative major subunit that is prone to rapidly polymerize prematurely into proteinaceous extensions in the bacterial cytoplasm when over expressed alone [115, 120]. Not only is this potentially toxic for the bacteria, but it also prevents proper T3SS biogenesis as these polymers are simply too large for secretion. To solve this problem, T3S chaperones such as CesAB [114, 115] or the co-chaperone family of PscE and PscG 120-123 are employed to trap the cognate subunits in a monomeric state prior to secretion.

Structural modeling has been used to predict how these T3S chaperones may actually prevent this subunit polymerization. In particular, structural data from the non-polymerizable needle monomer derivatives PrgI (*S. enterica* Typhimurium) [178], MxiH (*Shigella flexneri*) [179] and BsaL (*Burkholderia pseudomallei*) [180] was merged with the structural data of PscF needle subunits in complex with the PscE-PscG co-chaperone family (see Figure 3B) [122, 123]. This showed that by binding to a helical domain located in the needle subunit C-terminus, the PscG co-chaperone could prevent this domain from contributing sites for extensive surface contacts between polymerizing subunits [27]. It is still too early to tell whether this anti-polymerization model is broadly applicable to all T3SSs, especially since cognate T3S chaperones are not yet indentified for the needle proteins in most T3SSs. It is also important to establish the precise role for each of the co-chaperones. For example, is the PscG member solely responsible for anti-polymerization activity? Does the PscE derivative ensure co-chaperone stability and/or specific substrate piloting to the secretory apparatus? Whatever their individual roles, this model still nicely illustrates how T3S (co)chaperones can be used to quell the polymerization tendencies of highly aggregative proteins prior to transportation to their final assembly platform. In essence, this role is akin to a partitioning function ascribed to class II T3S chaperones.

Secretion Pilot for T3SS Specificity

T3S Promiscuity - A Conundrum

T3SSs are generally quite promiscuous in that a given substrate belonging to one system can readily be secreted by another heterologous system. This secretion promiscuity need not be restricted to genetically related systems. For example, effectors of animal pathogens are efficiently secreted by phytopathogens even though the latter produce a morphologically distinct T3SS [181, 182]. In addition, primordial T3SSs associated with the flagella biosynthesis pathway possess the ability to secrete, but not translocate, type III substrates associated with bacterial virulence [183-186]. This problem is compounded by the fact that a single genome can encode multiple independent T3SSs: those promoting motility associated with either polar flagella or lateral flagella, and those associated with eukaryotic cell contact (which can often exceed one system per genome) [10.] Because all these systems can be produced in the one bacterium, mechanisms must be in place to insulate each T3SS and ensure that individual substrates are secreted through the cognate system. One solution could have been to append common secretion signal sequences to all substrates of the same system; those with one particular signal are directed to one system, while those with another are directed to a different system and so on and so forth. This concept has been described previously for substrates of two different T3SSs produced by *S. enterica* Typhimurium (the SPI-1 and SPI-2 systems) [187], but in light of T3S promiscuity it has not gained wide acceptance.

On the other hand, in response to differing environmental cues it is clear that bacteria can isolate the expression of one system from all others. For example, *S. enterica* Typhimurium encodes two virulence associated T3SSs with distinct functions – one is encoded on the *Salmonella* pathogenicity island SPI-1 and is necessary for bacterial invasion into eukaryotic cells; the second on SPI-2 and functions in bacterial intracellular survival [188, 189]. Responding to complex, but distinct, extracellular and intracellular environmental cues, the expression of SPI-1 and SPI-2 components can be inversely coordinated by an interplay between various global and specific gene regulators [190]. This ensures that products promoting invasion are induced while products promoting intracellular survival are repressed and vice versa. In addition, these regulatory mechanisms also coordinate effector substrate production so that high levels are available for secretion at the same time as their cognate T3SS is active.

T3S Chaperones Confer Substrate Secretion-Pathway Specificity

However, another decisive means for ensuring substrate secretion-pathway specificity concerns an activity of T3S chaperones. As stated earlier, adjacent to the extreme N-terminal secretion signal of individual effector substrates lies about 100 amino acids that are essential for specific binding of their cognate chaperone – the CBD (Figure 3A). Mutagenesis within this CBD of *Salmonella* SPI-1 effectors, SptP and SopE, not only prevents respective SicP and InvB chaperone binding, but also leads to substrate secretion via the non-cognate T3SS of flagella 191, 192. This demonstrates that the presence of CBDs in the effector N-terminal region and subsequent binding by the cognate chaperone is required to guarantee that substrates are secreted through the correct T3SS. Importantly, this is not just a phenomenon of T3S chaperones of the effector class. Our own analysis revealed that in the absence of the LcrH chaperone, *Y. pseudotuberculosis* can secrete the YopB and YopD translocator proteins independent of a functional cognate Ysc-Yop T3SS; secretion probably occurs via another independent T3SS [Edqvist *et al.,* unpublished data]. Moreover, truncated versions of SipB translocator from *S. enterica* Typhimurium, which are unable to bind cognate chaperone, are also secreted via the flagella T3S, but not the SPI-1 T3SS [146]. T3S chaperones must therefore confer secretion pathway specificity, which prevents type III substrates associated with host-cell contact from being secreted by the flagella system, which might normally function as the default T3S pathway.

This suggests that the substrate-T3S chaperone complex must present a structural motif specifically recognized by the cognate T3SS. In the absence of either chaperone or the CBD within the substrate, secretion specificity is lost allowing secretion through the flagella pathway. Presumably, flagella-mediated secretion is supported by an extreme N-terminal secretion signal within the first 20 amino acids that represents a generalized canonical secretion signal used by all T3SS substrates – otherwise known as a chaperone-independent secretion signal (see below and Figure 3A). In contrast, the substrate-chaperone complex represents a second, unique secretion signal that aligns substrates for secretion via specific non-flagella T3SSs. As mentioned already in this chapter, several class I T3S chaperones have either been crystallized alone or in complex with their cognate substrate. Close scrutiny of multiple primary amino acid sequences and tertiary structures uncovered a conserved hydrophobic binding pocket in these chaperones that specifically accommodated a complementary short peptide of spatially conserved hydrophobic resides found within the chaperone binding domain of bound cognate substrates [98]. In light of this finding, it was suggested that many different T3SS substrates and cognate chaperones utilize this type of universal receptor-ligand interaction [98]. This structural commonality might well be the molecular mechanism governing secretion-pathway specificity that ensures the appropriate delivery of individual substrates to their cognate T3SS associated with bacteria-host cell contact. It is also otherwise known as a chaperone-dependent secretion signal. How this targeting mechanism may contribute to substrate secretion is discussed later.

Orchestrating Hierarchal Secretion

Prioritizing Early Substrate Secretion – The Specificity Switch

T3SSs are able to secrete three types of substrates; needle associated proteins and anti-activators (early substrates), translocator proteins (middle substrates) and effector proteins (late substrates). Following needle assembly completion (or the hook-basalbody in flagella biogenesis), a wealth of evidence supports a switching mechanism that triggers a change in substrate secretion from 'early' needle components (or flagella hook components) to the 'middle' translocator and 'late' effector proteins (or the flagella filament proteins). This is based on the crosstalk between the YscU-like, YscP-like and YscI-like proteins [193-199]. The inter-related sensing function of these protein families is universally conserved in both flagella and non-flagella T3SSs. The interested reader is referred to the following reviews for details of the complex molecular memchanism [13, 200-203]. Despite this clarity, comparatively little is known about a secretion order that may exist among the middle and late secretion substrates. Current dogma stipulates however that en-route to the eukaryotic cell interior, effector toxins pass through translocon pores in the host-cell plasma membrane that are composed of the translocator proteins [17, 26, 27]. For assembly of a functional type III translocon, this therefore implies that translocators must be secreted before effectors.

Creating Secretion Order among Middle and Late Substrates – T3S Chaperones and Beyond

Laboratory based secretion assays on fully induced T3SSs reveal no temporal differences between middle and late secretion substrates. Contributing to this is the asynchronous nature of T3SS assembly on the bacterial cell surface during in vitro culturing. At any time point, a bacterial cell will harbor both 'immature' T3SSs capable of secreting only early substrates and 'mature' T3SSs competent of secreting both middle and late substrates. This variable output thereby disguises any hint of ordered secretion. However, recent genetic evidence has provided the first indications that middle substrates are prioritized for secretion over late substrates. A heterogeneous family of proteins headlined by InvE of *S. enterica* Typhimurium influence T3SS substrate specificity by exhibiting a range of functions that essentially either promote the export of translocator (middle) substrates and/or inhibit the export of effector (late) substrates [204-212]. How these proteins function to distinguish translocators from effectors can vary. For example, InvE directly recognizes translocator-chaperone complexes that may prioritize their secretion [206]. Alternatively, the C-terminus of SepL may specifically bind effector substrates to stall their T3S from *E. coli* [204] or MxiC may bind the system ATPase creating a blockade that similarly inhibits effector secretion by *S. flexneri* [205]. No matter how it is achieved, these studies all represent further examples of built-in mechanisms for orchestrating hierarchal secretion among the T3S translocator and effector substrates.

T3S chaperones could also assist in establishing a secretion hierarchy. This idea was born from the initial discovery that only in the presence of other T3S substrates did the translocation of YopE into infected eukaryotic cells require both the cognate SycE chaperone and the internal CBD in YopE [213]. This suggested that SycE and the CBD conferred upon YopE a competitive advantage for secretion when in the presence of other substrates [213]. This has been reinforced by the proposal that substrates with two N-terminal secretion signals – the chaperone-independent and chaperone-dependent signals (see below) – might be prioritized for secretion over those substrates with no apparent chaperone-dependent signal [139]. As discussed in the following section, this is consistent with the perceived targeting role of chaperones on the back of their ability to bind to the ATPase, a core constituent of the secretory apparatus[98, 214-216]. In turn we have proffered that the obvious structural demarcation between class I and class II T3S chaperones could present two distinct recognition motifs to the ATPase as a mechanistic basis for establishing an order to secretion of the middle and late substrate categories [Amer *et al.,* unpublished data] [104]. However, a caveat in this concept is that so far no class II T3S chaperone is known to bind to the T3SS ATPase. Although this is not proof that an interaction does not occur, it is possible that recognition might be mediated through an additional component located at, or near, the portal for substrate entry into the secretion nanomachine. For example, the InvE protein is known to physically interact with translocator-chaperone complexes to selectively promote their secretion prior to the late effector substrates [206]. In addition, the inner membrane proteins InvI (from *S. enterica* Typhimurium SPI-1) and YscO (from *Y. enterocolitica*) may contribute to selective secretion of middle 'translocator' substrates before late 'effector' substrates [161]. These proteins actively bind to free class II, but not class I T3S chaperones at the cytoplasmic membrane following secretion of their translocator substrates through the T3SS. The idea is then that InvI/YscO function to recycle these 'empty' chaperones back to new substrates waiting in the secretion cue advancing their secretion more rapidly [161].

It is also worth noting that T3S chaperones establish secretion hierarchy through control of gene transcription. However, a discussion of this function will immediately follow below. In addition, it is worthwhile to highlight how regulation by proteolysis may influence temporal substrate secretion. In the phytopathogen *P. syringe,* the Lon protease effectively degrades T3S effector substrates, but only in the absence of their cognate chaperone [170]. Since Lon activity can be suppressed by T3S chaperones, they could theoretically implement a secretion order among the effectors based upon the degree of this suppression [170]. One can imagine that a chaperone providing significant protection from Lon-mediated degradation will leave more substrate available for secretion and vice versa. Significantly, powerful advances in fluorescent tagging techniques combined with real-time imaging offer unique opportunities to analyze spatial and temporal dynamics of type III substrate secretion. Already, these assays have proven sufficiently sensitive to detect hierarchal effector translocation into infected eukaryotic cells [217-220]. Therefore, they should also enable further exploration into the role of proteolysis as a mechanism for generating type III substrate secretion hierarchies.

Coupling of Secretion with System Regulation

Type III gene transcription is often coordinated through the activity of AraC-like transcriptional activators [44, 45, 221]. In addition, T3SSs are frequently subject to feedback inhibition such that secretion competence is normally required for expression of type III genes. An active T3S is sensed through the cytoplasmic accumulation of free 'co-activator' and the concomitant depletion of a secreted 'anti-coactivator'.

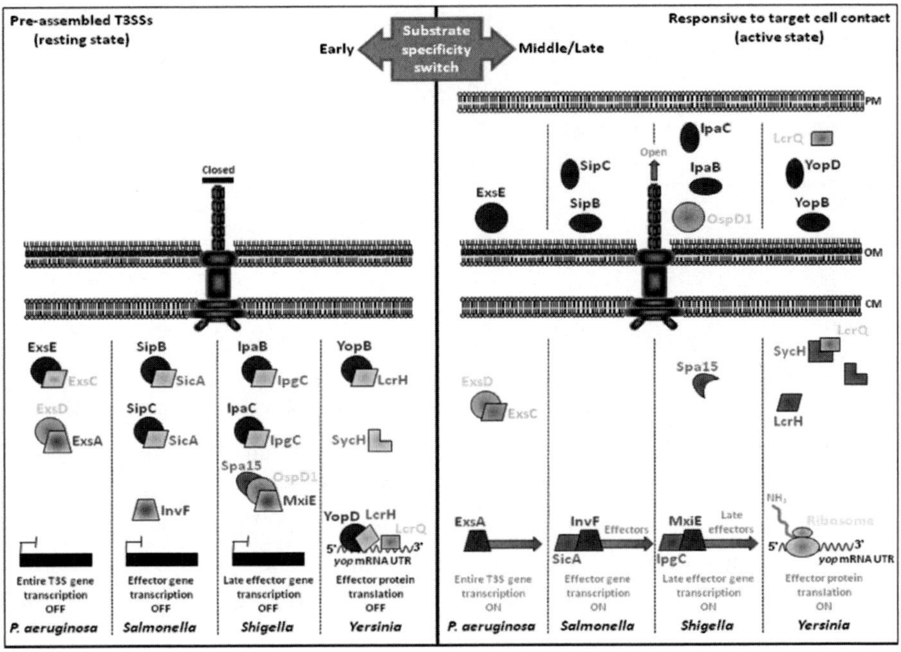

Figure 6. Schematic diagram summarizing the involvement of T3S chaperones in regulating T3SS activity in the animal pathogens Pseudomonas aeruginosa, Salmonella enterica, Shigella spp. and Yersinia spp. Early substrates comprising needle associated proteins are secreted first and assembled into a needle complex. During which time, production of late secreted substrates is kept in check by the action of T3S chaperones. In P. aeruginosa, the anti-anti-activator function of the ExsC chaperone is guarded by non-secreted pools of cognate substrate (ExsE), while the anti-activator ExsD silences the transcriptional activator ExsA. In S. enterica and Shigella, the co-activator T3S chaperones SicA and IpgC are inhibited by bound substrate (SipB/C and IpaB/C respectively). In Shigella, the anti-activator complex of Spa15/OspD1 further refines transcriptional repression of gene expression by binding to the transcriptional activator MxiE. An alternative post-transcriptional inhibitory mechanism in Yersinia relies on the anti-activator function of the LcrH T3S chaperone, which acts in concert with cognate substrate YopD and another regulatory element LcrQ to bind untranslated T3S gene mRNA. Following formation of a syringe-needle complex, the middle/late substrates are actively secreted following induction via target cell contact. This frees up ExsC to titrate away ExsD from EsxA (P. aeruginosa) or SicA/IpgC to directly activate InvF/MxiE (S. enterica/Shigella) and induce T3S gene transcription. By comparison, destabilization of the LcrH-YopD-LcrQ complex via chaperone (LcrH and SycH)-mediated substrate secretion (YopD and LcrQ, respectively) permits ribosome access and mRNA translation. This schematic was adapted from the recent review by Brutinel and Yahr 44. PM – plasma membrane (eukaryotic cell), OM – outer membrane, CM – cytoplasmic membrane.

Such information is then conveyed to an AraC-like transcriptional activator that establishes a cascade of type III gene activation. With differing degrees of complexity, variations of this general theme are built into several T3SSs. In every case however, there is reliance upon the ability of a T3S chaperone 'co-activator' to target for secretion its cognate secreted substrate 'anti-coactivator'. Examples of this T3S chaperone-dependent mode of system regulation are illustrated in Figure 6.

Pseudomonas Aeruginosa T3SS

The most recent example of this concept comes from elegant work with the *P. aeruginosa* T3SS. In this system, transcription of all genes is under the transcriptional control of ExsA, an AraC-like transcriptional activator. ExsA activity is restricted by three additional proteins with functions linked to the secretion competence of the T3SS. The first is an 'anti-activator' ExsD that binds to and inactivates ExsA when the T3SS is closed [222]. A closed T3SS also accumulates high cytoplasmic concentrations of ExsE enabling sequestration of an 'anti-anti-activator' ExsC [223, 224], a protein with class I chaperone characteristics 225. However, opening of the T3SS channel through target cell contact first allows ExsC to promote the secretion and translocation of ExsE reducing these internal inhibitory pools [223, 224, 226]. It then allows ExsC to seize ExsD, which in turn releases ExsA to activate transcription of type III gene promoters [225, 227].

Salmonella SPI-1 and Shigella T3SS

In contrast to this *P. aeruginosa* scenario in which the entire T3SS is up-regulated in response to activated secretion, T3SS activation in both *Salmonella* and *Shigella* only concerns the induction of effector genes. This illustration is based on the function of the translocator class II chaperones IpgC of *Shigella* and SicA of *Salmonella,* in concert with their two translocator substrates IpaB/IpaC and SipB/SipC respectively. In the absence of substrate, free IpgC or SicA acts in cooperation with the AraC-like transcriptional activators, MxiE and InvF respectively, to specifically transcribe mRNA from effector gene promoters [112, 163, 228-231]. In contrast, gene transcription is not induced when IpgC or SicA are bound to their cognate substrates trapped in the cytoplasm – therefore denying complex formation with their respective transcriptional activators. Hence, only after the secretion of their translocator substrates ('anti-coactivators') through an opened T3SS, is the chaperone free to act as a 'coactivator'. Notably, this regulatory coupling also doubles as a means to establish a secretion hierarchy because chaperone-activator complexes form and then trigger transcription of effector genes only after the cognate translocator substrates have been synthesized and secreted [229]. This mechanism, based on using a class II T3S chaperone as a coactivator, might be a common solution for creating a program to order type III substrate secretion by bacteria. For example, the chromosomal encoded T3SS of *Y. enterocolitica* depends upon a similar SycB chaperone/YsaE activator complex for transcriptional regulatory control, although in this case it is too early to say if this is coupled to secretion activity [232].

In *Shigella,* another layer of regulation is imposed to further refine this hierarchal secretion process. This is based upon an 'anti-activator' function to keep MxiE quiet, even in the presence of free coactivator IpgC. In this scenario, the anti-activator OspD1 (a secreted effector) would bind to MxiE shielding it from coactivation by IpgC. Only after secretion of OspD1, presumably mediated by the class Ib chaperone Spa15, will MxiE-IpgC complexes form to activate late effector genes that are under MxiE control [233]. This then assumes that fine-tuning of MxiE activation by the opposing anti-activator and co-activator regulatory pathways ensures late effectors are produced only after the translocators and early effectors (those not controlled by MxiE, such as OspD1) have been secreted through an active T3SS [233].

Yersinia T3SS

The previous examples describe systems which couple T3S activity to the control of gene expression at the transcriptional level. In contrast, *Yersinia* spp. combine activity of their plasmid encoded T3SS with control of gene expression using a post-transcriptional mechanism. In this case, relief from T3SS feedback inhibition comes from the secretion of at least one negative regulatory element, LcrQ [16, 234-236]. In the cytoplasm, LcrQ is thought to bind to the 5'-untranslated region of messenger RNA [237]. Upon induction through target cell contact, an activated T3SS releases LcrQ with the assistance of the class I chaperone SycH [238] Presumably, ribosomes then have access to translate each transcript, resulting in an upsurge of substrate synthesis and secretion [16, 236, 238, 239]. A complex of the YopD translocator and the LcrH class II chaperone is also a negative regulatory feature of T3S control in *Yersinia* [160, 240, 241]. It seems that YopD-LcrH complexes also bind to the 5'-untranslated region of mRNA collaborating with LcrQ in the post-transcriptional silencing of type III gene expression [Costa *et al.,* unpublished data] [237, 240, 241]. Although it is not yet apparent how a YopD-LcrH complex communicates with LcrQ, this model offers exciting possibilities for further investigation. It is important to clarify the actual targets of post-transcriptional control – is it restricted to the expression of genes encoding just the secreted substrates or does the regulon also include genes encoding the structural components? The current model also assumes that YopD is secreted along with LcrQ once the T3SS is activated. If so, this could free-up LcrH to initiate additional known contacts with other noted regulatory components (such as YscY, YscE and TyeA) [Francis, unpublished data] [158-160]. It remains to be seen how these other LcrH complexes fit into the regulatory scaffolds that control T3S gene expression in *Yersinia.*

LcrQ is a regulatory molecule unique to the *Yersinia* T3SS. It can reportedly interact with several T3S chaperones – most notably, the class I chaperones SycH and SycO and the class II chaperone LcrH [111, 159, 238, 242]. Intriguingly, another target of LcrQ was recently shown to be phosphoenol pyruvate carboxylase (PEPC) [243]. The consequence of this interaction was inhibition of PEPC activity, effectively reducing the biosynthesis of oxaloacetate derived amino acids. With these observations, LcrQ was suggested to be an important link between *Yersinia* T3SS and central metabolism; the dual ability of LcrQ to bind both T3S chaperones and PEPC, might ensure the availability of amino acids at times when the synthesis of T3S substrates are most needed – during target cell contact [243]. The model proposes that following T3S of premade pools of substrate, LcrQ is sequestered by free T3S chaperone. This liberates PEPC to re-establish amino acid pools to initiate a subsequent round of T3S substrate synthesis. To ensure their solubility and stability prior to secretion, these de novo synthesized substrates recruit their cognate T3S chaperone. This releases LcrQ to seek out and neutralize PEPC activity to reduce unnecessary synthesis of amino acids. Since earlier studies have hinted at a link between metabolic status and regulation of flagella and non-flagella T3SSs [244-250], this cross-talk model offers a step forward and therefore deserves to be rigorously tested. At the very least, it stands to provide important insight into the relevance of the varied LcrQ-T3S chaperone interactions in *Yersinia.*

Variations on the Same Theme

The regulatory mechanisms described herein are ingenious ways in which bacteria have coupled their secretory activity to the control of gene expression. The key mediators of these regulatory pathways are chaperones or chaperone-assisted secreted substrates. They sense the operational 'readiness' of each T3SS and relay this information to transcriptional activators that sit at or near the top of this regulatory cascade. As already noted by others [44], a feature of these regulatory mechanisms are the inherent flexibility of the individual components. A good illustration of this is the varied functions of the class II T3S chaperones. The *Salmonella* and *Shigella* systems utilize their respective SicA and IpgC chaperones as cofactors for the transcriptional activator, while the *Yersinia* system employs the equivalent LcrH chaperone as a post-transcriptional negative regulator with no known connection to the transcriptional activator [Carlsson *et al.*, unpublished data]. On the other hand, the *P. aeruginosa* system does not engage the homologous PcrH in any apparent regulatory activity [158, 162]. Such resourcefulness probably reflects how related bacteria have utilized different environmental conditions to ensure type III genes are expressed only when needed (for example; during target cell contact within a unique infection niche), whereas energetically wasteful expression is kept to a minimum.

Chapter 5

Mechanisms of Substrate Secretion

The means by which T3S substrates are delivered to the cognate T3SS and then subsequently secreted continues to be the focus of intense research in the T3S field. Insights into several inherently complex mechanisms have been revealed. While they all contribute towards efficient substrate secretion, the various molecular details are still poorly understood. Adding to this complexity is the knowledge that the portal through which the substrates are assumed to be secreted, the needle channel, is far too narrow to accommodate folded proteins [27, 251.] Hence, energetically favorable means to unfold substrates is also a prerequisite for secretion [252].

Chaperone-Independent N-Terminal Secretion Signals

It was realized early on that the translocated effector substrates of *Yersinia* contained information within their first 20 amino acids that is sufficient to confer upon signalless reporter proteins an ability to be secreted in a type III-dependent manner[253, 254]. Subsequent in depth analysis of effector substrate N-termini failed to identify any consensus amino acid sequence; although a contributing amphipathic secondary structure was proposed [171, 255]. Secretion of some effector substrates is also aided by information in the 5'-prime untranslated mRNA sequence [256, 257]. However, some other effector substrates apparently contain a N-terminal secretion signal devoid of any secondary structure in either amino acid or mRNA signals [258]. We and others have also tried to resolve the precise structural features of the translocator substrate N-terminal secretion signals. Although the molecular makeup is still uncertain, it seems unlikely that mRNA signals contribute to translocator secretion [Amer *et al.*, unpublished data] [145, 146, 259]. Nevertheless, evidence supporting the mRNA and protein sequence models has intriguing consequences, for each proposes a different mechanism of substrate recognition and subsequent secretion by the T3SS. The mRNA hypothesis implies that a mRNA structure is first recognized by the T3SS that then allows coupling of translation directly to secretion (Figure 7A). On the other hand, the protein hypothesis implies recognition of pre-made pools of the N-terminal polypeptide (Figure 7B). It is clear that the N-terminal secretion-signal debate is not over. However, it may never really be resolved when most studies examine substrate secretion in artificial conditions that may not reflect the true inducing signal – target cell contact. More detailed information concerning the function of this so-called chaperone-independent N-terminal secretion signal can be obtained from several worthy reviews [13, 260-263].

Chaperone-Dependent Secretion Signals

As discussed earlier, located directly adjacent to the N-terminal secretion signal are about 100 amino acids of substrate that are involved in direct binding to the cognate chaperone.

It is now clear that chaperone bound to this region comprises a chaperone-dependent secretion signal [264]. This has consequences for prioritizing T3S when in competition with other substrates [213] and in ensuring specific secretion through the correct T3SS pathway [191, 192]. Secretion pathway specificity is conferred through association of chaperone-substrate complexes with an ATPase linked to the cytoplasmic side of the T3S machinery (Figure 7C). This was first examined in enteropathogenic *Escherichia coli* for the T3S substrate Tir and its chaperone CesT and their interactions with the T3SS ATPase EscN [214, 215]. Similarly, within the *S. enterica* Typhimurium SPI-1 T3SS, the InvC ATPase interacts with the SptP effector substrate in complex with its chaperone SicP [216]. Such findings endowed T3S chaperones with a new responsibility – targeting their substrate cargo to the cognate T3SS through specific associations with the core ATPase.

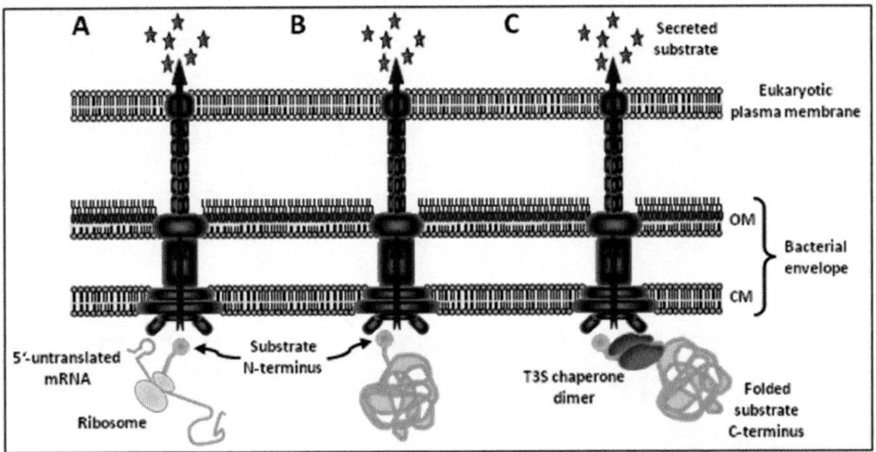

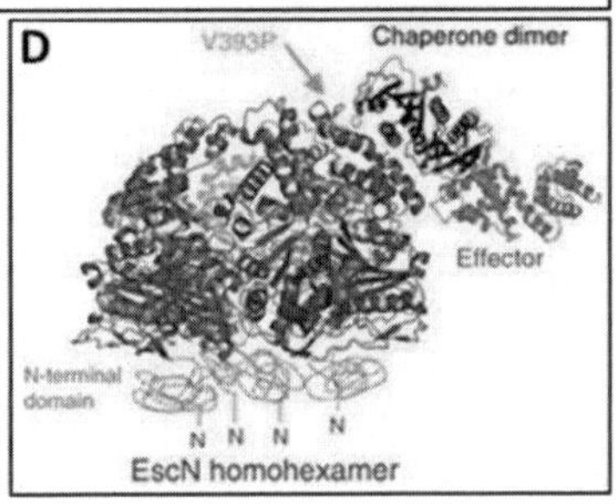

Figure 7. A schematic illustration of the possible mechanisms of type III substrate secretion. The T3SS may recognize the 5'-untranslated mRNA, with the implication that this recruits ribosomes to the inner face of the secretion apparatus and couples substrate translation to secretion (A). The amino acid sequence of the extreme N-terminus of individual substrates may represent a structural motif directly recognized by the T3SS (B). Information contained in the T3S chaperone bound to substrate may also constitute a secretion signal. It is the chaperone that interacts with the T3SS (most likely the system ATPase) and this primes the substrate for unfolding prior to secretion (C). This schematic was motivated by reviews from Aldridge and Hughes 404 and Cornelis 263. The chaperone dimer (red) in complex with its cognate substrate (green) is proposed to dock at a specific site of the T3SS ATPase (blue) that encompasses the functionally important valine (V; marked in yellow) residue at position 393 (D). Panel D is part of a study by Strynadka and colleagues 273 and is reused with permission "Copyright (2007) Macmillan Publishers Ltd." OM – outer membrane, CM – cytoplasmic membrane. OM – outer membrane, CM – cytoplasmic membrane.

This does not offset the role of the extreme N-terminal chaperone-independent secretion signal, because in some experimental setups, chaperone-free effector substrate could still associate with their respective system ATPase

[214, 265]. Thus, chaperone–dependent and –independent substrate secretion signals probably act in concert to dock the substrate to the T3S apparatus. For reasons that are outlined below, the T3SS ATPase is an appealing target for docking of substrate at the inner face of the T3S machinery. However, it is prudent to remain open to other possible interactions given that studies in *Yersinia* suggest that the ATPase YscN does not engage free SycE chaperone or a SycE-YopE effector complex [140]

Substrate-Chaperone Recognition by the T3S Machinery

The T3SS ATPase is ubiquitous to all T3SSs. The family typically contains the two Walker boxes, A and B, which are associated with ATP binding and hydrolysis. In addition, all exhibit some similarity to the β-subunit of F0F1 proton-translocating ATPases. Consistent with these observations, T3SS ATPases hydrolyze ATP in line with their role as energizers for substrate secretion [266-271]. The ATPases form a hexameric ring-like structure that probably sits at the base of the T3S machinery peripherally associated with the inner membrane [272, 273]. Oligomerization is a known stimulator of ATPase activity and might be critical for driving substrate secretion [269, 270]. However, other T3SS components are presumed to interact with the ATPase, which might be important in regulating oligomeric formation or perhaps even in restricting ATPase activity to the process of T3S [268, 270, 271, 274, 275].

So what is the true function of these ATPases? The crystallization of various T3SS components revealed a puzzling conundrum – fully folded secreted substrates are far too large to be accommodated by the small opening of the T3S nanomachine [27, 251, 276]. For secretion to occur, substrates must therefore assume an unfolded state. As mentioned earlier, T3S chaperones cannot initiate substrate unfolding. While the region of each substrate contacting the chaperone is unfolded, the remainder of the substrate is usually still fully folded [94, 103, 139, 173]. However, the findings that T3S chaperones dock their substrates at the T3SS ATPase proffered an attractive resolution – ATPase activity initiates substrate unfolding prior to transit through the T3S machine. In a seminal study of T3S in *S. enterica* Typhimurium, not only did the ATP hydrolyzing activity of the InvC ATPase disassociate the SptP effector from the SicP chaperone-bound state, but it also caused the effector to unfold 216. In this context, T3S chaperones pilot their cargo to the system ATPase, which then triggers chaperone-substrate disassembly and substrate unfolding that facilitates substrate entry into the secretory apparatus. Structural determination of the enteropathogenic *E. coli* T3S ATPase EscN, revealed a surface cavity in the C-terminus that could serve as a substrate-chaperone binding site (Figure 7D) [273]. ATP hydrolysis could induce structural changes in this region that precipitates chaperone disassociation [273]. However, ATPase-dependent unfolding must also be assisted by unknown structural properties inherent to T3S substrates because when fused to tightly folded heterologous protein domains, substrate unfolding and secretion is severely impaired [216, 277-279]. Thus, only substrates permissive to ATPase-dependent unfolding are secretion competent [216].

The energy to drive the substrates through the T3S machinery remains an open question. Work from the Galán laboratory demonstrates the importance of ATP hydrolysis in substrate unfolding [216]. In addition, some observations suggest that the proton motive force may contribute to the passage of substrate through the secretion channel [201, 252, 280]. Significantly, aside from their secretion piloting role, T3S chaperones are also hypothesized to reduce the energy requirements needed for substrate secretion. The majority of structural evidence indicates that when in complex with chaperone, the CBDs of substrates are maintained in an extended, non-globular conformation (see Figure 3B) [94, 98-100, 103, 109, 139]. Maintenance of this localized unfolded state by a bound chaperone may act as a primer generating a more energetically favorable environment for further substrate unfolding by the ATPase during presentation of the substrate to the secretion apparatus [86, 98, 99, 279, 281]. In contrast to this model however, binding by the SycE chaperone to the full-length YopE effector actually induced a localized disordered-to-ordered structural transition in the substrate CBD [140]. While these findings may not agree with the chaperone 'priming' model, they reaffirm that chaperones probably contribute to a recognition signal that directs their substrate to the T3S secretory machine.

Chapter 6

T3S Chaperones for Protein Export in Flagella Assembly

Bacterial swimming motility is conferred by the rotation of a complex nano-motor termed the flagellum. These filamentous organelles extend from the bacterial inner membrane to well beyond the bacterial surface. A flagellum is composed of three parts; the basal body entrenched in the bacterial envelope (the motor), followed by a hook (a connecting 'universal joint') and then a filament (the propeller) which protrudes outwards from the bacterial surface [201, 282, 283]. Remarkably, the structure of the flagellum hook-basal body closely resembles non-flagella T3SSs associated with host cell contact (Figure 1) [198, 251, 284-289]. This adds further impetus to the idea that T3SSs needed for interactions between bacteria and eukaryotic cells are descendents of the protein export system linked to bacterial flagella assembly [290]. The flagella-associated T3SS exports a number of proteins needed for flagella assembly [282]. At least four of these are subunits of the flagellum filament, which assemble outside of the bacterium; FliC (major filament subunit), FliD (filament cap) and FlgK and FlgL (the hook-filament junction proteins). Export of each of these proteins requires assistance from a dedicated T3S chaperone; FliS (for FliC) [291-293], FliT (for FliD) [294, 295] and FlgN (for FlgK and FlgL) [294, 295].

Mediators of Anti-Polymerization

Despite physical similarities to class I T3S chaperones of host cell contact inducible T3SS – being small, predominantly α-helical with an acidic pI –

flagella-specific T3S chaperones display notable differences warranting their division into a different T3S chaperone group, designated class III (Table 1) [88]. This is reinforced by a FliS crystal structure that is unrelated to any other T3S chaperone (Figure 2) [296]. Intriguingly however, a similar means of substrate recognition and localized substrate unfolding may still apply – the FliC substrate wraps around the outside of FliS in an extended confirmation lacking any teriary structure (Figure 3B) [296]. In addition, flagella class III T3S chaperones bind to the extreme C-terminus of their substrates [291-293, 295]. These C-terminal domains are thought to be the glue by which individual filament subunits polymerize together during assembly of a growing flagellum. The monomeric forms are therefore inclined to rapidly oligomerize unless the C-terminal interacting domains are stabilized by a bound chaperone. This has been experimentally demonstrated for the FliS chaperone [291-293 and the FlgN chaperone [294]; their respective binding to the major flagellum filament subunit FliC and the hook associated axial protein FlgK prevents premature polymerization. This anti-polymerization role is analogous to the class IV and class V T3S chaperones belonging to T3SSs associated with host cell contact [115].

Establishing Ordered Substrate Secretion

Preventing polymerization of cognate flagella substrates is arguably the foremost role of flagella-specific T3S chaperones. However, assembly of filament subunits into the growing flagellum is strictly sequential; the hook-filament junction proteins, FlgK and FlgL, must be assembled before the FliD filament cap, which in turn is needed for FliC flagellin filament incorporation [201, 282, 283]. This implies a strict secretion order, which may result if each chaperone-substrate complex interacts with a different affinity to a component of the flagella-specific T3SS. On top of this, competition for recognition of the FlgK, FlgL and FliD minor flagellum subunits by the T3SS would need to occur in an unfavorable environment replete with excessive amounts of FliC major flagellum subunit [282, 297]. Investigation into this problem by the Colin Hughes laboratory revealed two key observations. First, FlgN-FlgK and FlgN-FlgL chaperone-substrate complexes are recognized by the system ATPase FliI [298], a finding analogous to some class I T3S chaperones associated with virulence [214-216. Extrapolating from the work of Jorge Galán and colleagues [216], the flagella-specific FliI ATPase probably disassociates the FlgN-FlgK and FlgN-FlgL complexes to promote FlgK and FlgL unfolding and secretion. However, what happens to the 'free' T3S chaperone post-substrate secretion? This was investigated in a second study of Hughes and colleagues. They described an event post-substrate secretion in which free chaperone is then selectively recycled to another substrate to promote rapid export [297]. Significantly, the protein involved in this chaperone recycling process, FliJ, interacts only with FlgN and FliT, chaperones of the minor filament subunits. FliJ does not interact with the major filament chaperone FliS. Thus, by specifically recruiting newly available free FlgN and FliT chaperone and then quickly making it available to the next cognate substrate in line for secretion, FliJ-chaperone complexes may therefore confer an advantage towards the secretion of minor filament subunits over the FliC major subunit [297]. Further credence for such a novel hierarchal secretion mechanism comes from a preliminary analysis of two FliJ orthologues from cell contact inducible T3SSs, namely InvI (from *S. enterica* Typhimurium SPI-1) and YscO (from *Y. enterocolitica*) – both seemed to preferentially bind to free class II T3S chaperones targeting the translocators compared to class I T3S chaperones targeting the effectors [161]. Hence, class II T3S chaperone complexes with InvI or YscO might be expected to prioritize translocator substrate secretion. This work with the FliJ protein family has highlighted a new mechanistic concept for orchestrating hierarchal secretion, which warrants further study to determine its biological relevance.

Coordinating Gene Expression with Flagella Assembly

The coordinated assembly of flagella is a complex process – a review of these regulatory mechanisms within the flagella regulon is beyond the scope of this article. However, interested readers are referred to a collection of recent reviews that give a thorough description of this topic [282, 299-301]. In very brief terms, structural genes are expressed in a three-tiered transcriptional hierarchy based on class I, II and III promoter activity that are coupled to built-in assembly checkpoints within the growing flagellum. This transcriptional coordination ensures the sequential expression of class I genes (encoding the master regulators FlhC and FlhD for class II gene expression) before class II genes (encoding hook and basal body components as well as the σ28 sigma factor required for class III gene expression) before class III genes (encoding late secreted substrates and chemotaxis components). Each gene product is therefore made only when needed during the assembly process.

In this context, it is significant that all late secretion substrates – i.e., those which are secreted only after hook-basal body assembly is completed – require a chaperone [302]. Nor is it a coincidence that most of these flagella-class T3S chaperones (with the exception of FliS) exhibit intricate regulatory functions (Figure 8). These dual functional T3S chaperones therefore create the links between transcriptional regulation and sensing of flagellum assembly status. For example, the substrate FliD sequesters its cognate FliT T3S chaperone in the cytoplasm prior to completion of a secretion competent hook-basal body [303]. Once completed however, FliD is escorted by FliT for secretion through the hook-basal body. The FliT chaperone is then free to complex with the FlhC subunit of the FlhC-FlhD master regulator complex inhibiting its binding to class II promoter DNA [303]. This negative-regulatory loop is anticipated to fine tune class II gene expression, reducing the production of hook and basal body components following their assembly into a functionally complete structure. It also seems apparent that the FlgN T3S chaperone similarly contributes to the controlled translation of *flgM,* a class III gene encoding the anti-σ[28] factor[304, 305].

In this scenario, the substrates FlgK and FlgL sequester FlgN in the cytoplasm before the hook-basal body is completed [304]. When completed however, FlgK and FlgL are secreted through the hook-basal body freeing the FlgN chaperone to positively regulate *flgM* translation. These subtle alterations in the levels of FlgM are expected to have significant effects on σ28–dependent transcription of class III transcripts. The anti-σ28 factor FlgM also needs to be secreted following assembly of the hook-basal body structure. This is needed in order to free up the σ28 sigma factor to direct the RNA polymerase to class III promoters upstream of genes encoding late flagellar components [306]. Interestingly, the σ28 sigma factor is the T3S chaperone for its own anti-σ28 factor FlgM [302].

While flagella and non-flagella T3SSs have different pursuits – one to propel bacterial movement toward or away from a chemical stimulus, the other to establish cross-talk with eukaryotic cells – fundamental similarities between the two protein export systems exist. It is now evident that the role of T3S chaperones is universal among both flagella and non-flagella T3SSs, and their influence is observed at multiple levels. Although their structural scoffolds and modes of action are sometimes different, their motives are always the same – to protect substrates from premature associations with self or other proteins; to guide substrates to the correct T3SS; to establish a secretion order among the substrates;

and to couple gene transcription to the process of apparatus assembly (see Figure 5).

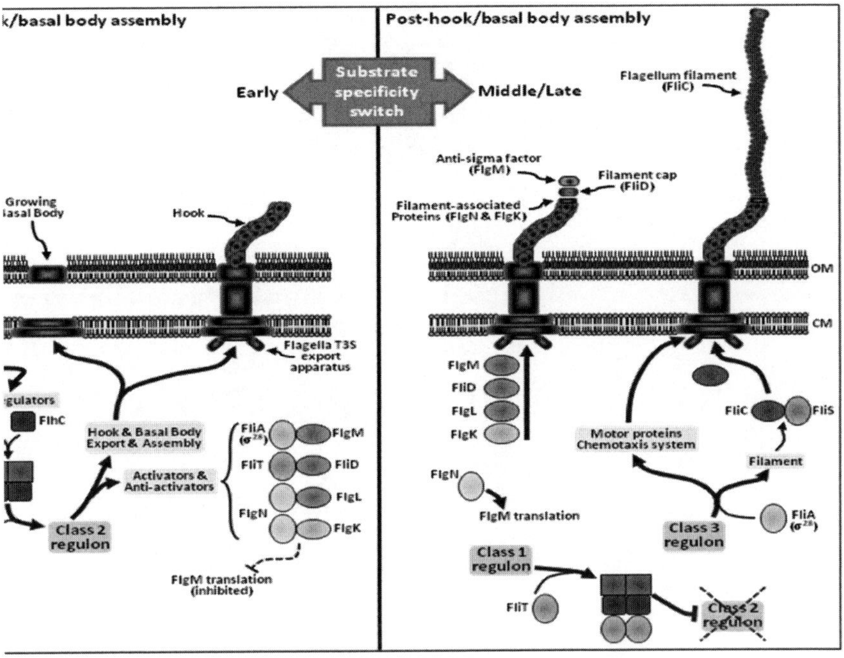

Figure 8. Use of chaperones to couple flagella gene expression with flagellum assembly. The initial stages of flagella biogenesis require the assembly in the bacterial envelope of a hook-basal body structure. Components of this structure are encoded on a series **of** σ^{70}-dependent operons within the Class II regulon, which is under the control of the master regulators (FlhC and FlhD) encoded within an operon of the Class I regulon. Late secretion substrates are also produced, but cannot be secreted until the hook-basal body is secreted. They are stabilized in the cytoplasm prior to secretion by bound T3S chaperones. Completion of the hook-basal body signals a switch in substrate specificity allowing for secretion of late substrates. Secretion of the anti-sigma factor FlgM frees-up the cognate **chaperone,** FliA (σ^{28}), to initiate expression of genes within the Class III regulon. This includes the major filament subunit FliC and its cognate chaperone FliS, which allows assembly of the external flagellum. In addition, when freed by the secretion of FlgL and FlgK, the FlgN chaperone positively regulates FlgM production. Moreover, free FliT can engage with FlhC of the master regulator complex inhibiting expression of genes from the Class II regulon. This schematic was inspired by the reviews of Chevance and Hughes [282] and Brutinel and Yahr [44] as well as primary studies from the Hughes [304] and Kutsukake [303] laboratories. OM – outer membrane, CM – cytoplasmic membrane.

Chapter 7

Chaperones of other Specialized Host-Targeting Protein Secretion Systems

In recent years and aided by the whole genome sequencing explosion, two new secretion systems designated Type IV (T4SS) and Type VI (T6SS) have been added to the bacterial protein export arsenal. Analogous to T3SS, these additional systems are also believed to boast a capacity to directly transport proteins from the bacterial cytosol into host cell compartments.

Type IV Substrate Secretion

Just like T3SSs are ancestrally related to the bacterial flagella, T4SSs are related to the bacterial conjugation system that transfers DNA from donor to recipient cells. [307-310] As such, various T4SSs can be used to import DNA, or secrete DNA, protein or DNA-protein complexes free into the environment or directly into target eukaryotic cells. The latter is a process that resembles T3S effector translocation. The list of so-called effector molecules directed to the host cell by T4SS's is growing rapidly. However, in most situations it is poorly understood how these molecules are targeted to the secretory apparatus and prepared for orderly and efficient secretion. Secretion signal information usually exists in the C-terminus of most substrates, but this is not always sufficient for secretion [310, 311]. Moreover, in almost all T4SSs the substrate recognition, recruitment and export also relies upon an integral inner membrane located receptor termed a 'coupling factor' [312]. Finally, some researchers propose that T4SSs incorporate chaperone-like molecules to perform functions akin to the T3S chaperones – preventing premature interactions of substrate with self or other proteins, maintaining substrates in a secretion competent state or delivering the substrates to the coupling protein [311].

With respect to the T4SS encoded by the plant pathogen *Agrobacterium tumifaciens*, one of the transported substrates VirE2 is stabilized by the VirE1 chaperone [313-317]. However, this interaction is not needed for transfer of substrate to the coupling protein and subsequent export [318, 319]. The chaperone therefore likely restricts premature VirE2 polymerization and binding to single stranded T-DNA in the bacterial cytoplasm. In this regard, it is interesting that although unique, the VirE2-VirE1 crystal structure does evoke some resemblance to the anti-polymerization complexes formed by the T3S chaperones FliS and CesAB with their respective substrates FliC and EspA (Figure 3C) [115, 296, 320].

The intracellular pathogen *Legionella pneumophila* also utilizes a T4SS for translocation of vast numbers of effectors into the host cell cytosol. The translocation of a subset of these effectors requires an interaction with the IcmS/IscW or IcmS/LvgA co-chaperone complexes in the bacterial cytoplasm [321-326]. However, how these co-chaperone complexes contribute to the substrate translocation process is poorly understood, but one role might be to present the C-terminal substrate secretion signal for recognition by the T4SS [321]. The *L. pneumophilia* IcmR protein also exhibits chaperone activity for IcmQ [327, 328]. Not only does chaperone binding prevent IcmQ self-association, but it also prevents IcmQ pore formation in phospholipid membranes [329, 330]. Intriguingly, this chaperone-substrate complex is functionally reminiscent of the VirE1-VirE2 complex, where VirE2 also forms channels in planar lipid bilayers [331]. Parallels can also be drawn with the class II T3S chaperones that prevent the damaging effects of premature membrane insertion by the T3S translocator substrates [Costa *et al.*, unpublished data] [142, 147].

The last example of a putative chaperone-like molecule among components of T4SSs comes from the *cag* system encoded by the stomach pathogen *Helicobacter pylori*. CagA is the only known proteinaceous T4S substrate secreted by *H. pylori*. Secretion first requires an interaction with CagF that is primarily located in the cytoplasmic membrane suggesting that this initial binding might contribute to subsequent recognition of CagA by the type IV secretory apparatus [332, 333].

Collectively, these studies do suggest a universal need for unique chaperone-like molecules in coordinating T4S in diverse bacteria. However, indications are that at least some T4S chaperones have mechanistically diverged from their T3S chaperone counterparts – clearly a reflection of the different molecular constraints imposed upon by the two evolutionary distinct secretion systems within which each chaperone family are designed to operate.

Type VI Substrate Secretion

Far less is known about the T6SS since its widespread discovery has only occurred within the last couple of years. However, its discovery has prompted significant interest in the research community seeing as it may yet be another important strategy for bacterial survival in the presence of eukaryotic cells [334-337]. The system can consist of up to 25 proteins, most of which lack homology to any protein belonging to other independent bacterial secretion systems. It follows that few secreted substrates have so far been reported. It is not clear how these substrates are secreted, or whether they require chaperone assistance for their stabilization and recognition by the secretory apparatus. The presence of dedicated chaperones unique to T6SSs has been speculated, but this is not supported by any direct experimental evidence [338, 339].

ClpV, a subfamily of the ClpB chaperone-like ATPases associated with maintenance of protein complexes, is most often associated T6SS [340]. However, it's chaperone capacity is considered secondary to its ATP binding and hydrolysis activity, which may function as the general energizer for type VI substrate secretion [341]. None the less, the role of ClpV is not a transparent one. While recent data confirms that the ClpV ATPase activity is essential for T6S, it might preferentially be needed to disassemble a tubular-like structure composed of two proteins encoded by genes within the T6SS cluster [342]. However, the function of this tubular-like structure in T6S is not known, which brings into question the significance of its disruption by ClpV. Additional studies should eventually reveal the precise relationship between ClpV activity and T6SS. Hopefully further studies will also determine if other cytoplasmic chaperone-like accessory proteins are required to assist with type VI substrate secretion.

Conclusion

Use of the term 'T3S chaperone' has been met with some trepidation and confusion. This occurs particularly among those familiar with the diverse family of molecular chaperones that usually function to prevent non-functional protein complexes by promoting correct protein folding and to encourage the refolding of misfolded proteins [343]. In recent times however, new concepts in T3S chaperone biology have emerged courtesy of several recent structure-function studies. For example, some T3S chaperones might primarily be sensors of T3SS assembly, while others could be facilitators of T3S export [221, 300, 305]. Thus, while the term 'chaperone' may not be entirely consistent with the classical view of molecular chaperone function, it still admirably fits into the concept that T3S chaperones are critical for maintaining and orchestrating the biomechanics of a well-oiled nanomachine that in turn, enables multiple bacteria to establish interactions of significance with host organisms. From humble beginnings it has quickly become apparent that these small molecules pack an impressive array of functions in their small molecular scaffold. Continued research in this field promises to reveal further examples of bacterial engineering ingenuity through the sculpturing of T3S chaperones with impressive versatility – effectively offering the use of a compact small molecular toolkit for all occasions.

In closing, just like molecular chaperones are emerging as new targets for drug therapy of microbial virulence [344, 345], the importance of T3S chaperones in bacterial pathogenicity presents us with a golden opportunity to exploit their functions in the search for novel and effective anti-bacterial drugs. With the emergence of crystal structures of T3S chaperones in complex with substrates, this structural data could enable the design of small synthetic molecules that specifically block T3S chaperone-substrate interaction via competitive antagonism. Such an approach is already showing promise for the targeted disruption of periplasmic chaperone binding to substrates that then assemble into surface pili appendages via the chaperone-usher protein secretion pathway 346. Another independent approach could be to randomly identify small inhibitory molecules of T3S chaperone function through the high throughput screening of large chemical libraries. Indeed, high throughput screening procedures are successfully identifying synthetic chemicals with inhibitory effects against bacterial virulence determinants including T3SSs [347, 348].

Acknowledgements

Thanks are extended to the numerous T3S researchers world-wide - their experimental and intellectual contributions have provided a continued impetus for greater understanding of T3S chaperone biology. Work in the Francis laboratory is performed within the framework of the Umeå Centre for Microbial Research and the Laboratory for Molecular Infection Medicine Sweden. Past and present grant support from the Swedish Research Council, Carl Tryggers Foundation for Scientific Research, Umeå University Basic Science-Oriented Biotechnology Research Fund, Foundation for Medical Research at Umeå University, Swedish Cystic Fibrosis Association Research Fund and Swedish Medical Association is gratefully acknowledged.

References

[1] Fauvart, M. and Michiels, J. (2008). Rhizobial secreted proteins as determinants of host specificity in the rhizobium-legume symbiosis. *FEMS Microbiol Lett, 285,* 1-9.

[2] Finlay, B. B. and Falkow, S. (1997). Common themes in microbial pathogenicity revisited. *Microbiol. Mol. Biol. Rev., 61,* 136-169.

[3] Dale, C. and Moran, N. A. (2006). Molecular interactions between bacterial symbionts and their hosts. *Cell, 126,* 453-465.

[4] Preston, G. M. (2007). Metropolitan microbes: type III secretion in multihost symbionts. *Cell Host Microbe, 2,* 291-294.

[5] Goebel, W. and Gross, R. (2001). Intracellular survival strategies of mutualistic and parasitic prokaryotes. *Trends Microbiol., 9,* 267-273.

[6] Soto, M. J., Dominguez-Ferreras, A., Perez-Mendoza, D., Sanjuan, J. and Olivares, J. (2009). Mutualism versus pathogenesis: the give-and-take in plant-bacteria interactions. *Cell Microbiol., 11,* 381-388.

[7] Ochman, H. and Moran, N. A. (2001). Genes lost and genes found: evolution of bacterial pathogenesis and symbiosis. *Science, 292,* 1096-1099.

[8] Hentschel, U., Steinert, M. and Hacker, J. (2000). Common molecular mechanisms of symbiosis and pathogenesis. *Trends Microbiol, 8,* 226-231.

[9] Hueck, C. J. (1998). Type III protein secretion systems in bacterial pathogens of animals and plants. *Microbiol. Mol. Biol. Rev., 62,* 379-433.

[10] Pallen, M. J., Beatson, S. A. and Bailey, C. M. (2005). Bioinformatics, genomics and evolution of non-flagellar type-III secretion systems: a Darwinian perpective. *FEMS Microbiol. Rev., 29,* 201-229.

[11] Troisfontaines, P. and Cornelis, G. R. (2005). Type III secretion: more systems than you think. *Physiology (Bethesda), 20,* 326-339.

[12] Lee, V. T. and Schneewind, O. (2001). Protein secretion and the pathogenesis of bacterial infections. *Genes Dev., 15,* 1725-1752.

[13] Galan, J. E. and Wolf-Watz, H. (2006). Protein delivery into eukaryotic cells by type III secretion machines. *Nature, 444,* 567-573.

[14] Kimbrough, T. G. and Miller, S. I. (2002). Assembly of the type III secretion needle complex of *Salmonella typhimurium. Microbes Infect., 4,* 75-82.

[15] Rosqvist, R., Magnusson, K. E. and Wolf-Watz, H. (1994). Target cell contact triggers expression and polarized transfer of *Yersinia* YopE cytotoxin into mammalian cells. *EMBO J., 13,* 964-972.

[16] Pettersson, J., Nordfelth, R., Dubinina, E., Bergman, T., Gustafsson, M., Magnusson, K. E., *et al.* (1996). Modulation of virulence factor expression by pathogen target cell contact. *Science, 273,* 1231-1233.

[17] Mueller, C. A., Broz, P. and Cornelis, G. R. (2008). The type III secretion system tip complex and translocon. *Mol. Microbiol., 68,* 1085-1095.

[18] Mueller, C. A., Broz, P., Muller, S. A., Ringler, P., Erne-Brand, F., Sorg, I., *et al.* (2005). The V-antigen of *Yersinia* forms a distinct structure at the tip of injectisome needles. *Science, 310,* 674-676.

[19] Espina, M., Olive, A. J., Kenjale, R., Moore, D. S., Ausar, S. F., Kaminski, R. W., *et al.* (2006). IpaD localizes to the tip of the type III secretion system needle of *Shigella flexneri. Infect. Immun., 74,* 4391-4400.

[20] Johnson, S., Roversi, P., Espina, M., Olive, A., Deane, J. E., Birket, S., *et al.* (2007). Self-chaperoning of the type III secretion system needle tip proteins IpaD and BipD. *J. Biol. Chem., 282,* 4035-4044.

[21] Sani, M., Botteaux, A., Parsot, C., Sansonetti, P., Boekema, E. J. and Allaoui, A. (2007). IpaD is localized at the tip of the *Shigella flexneri* type III secretion apparatus. *Biochim. Biophys. Acta, 1770,* 307-311.

[22] Broz, P., Mueller, C. A., Muller, S. A., Philippsen, A., Sorg, I., Engel, A., *et al.* (2007). Function and molecular architecture of the *Yersinia* injectisome tip complex. *Mol. Microbiol., 65,* 1311-1320.

[23] Jin, Q. and He, S. Y. (2001). Role of the Hrp pilus in type III protein secretion in *Pseudomonas syringae. Science, 294,* 2556-2558.

[24] Li, C. M., Brown, I., Mansfield, J., Stevens, C., Boureau, T., Romantschuk, M., *et al.* (2002). The Hrp pilus of *Pseudomonas syringae* elongates from its tip and acts as a conduit for translocation of the effector protein HrpZ. *EMBO J., 21,* 1909-1915.

[25] Crepin, V. F., Shaw, R., Abe, C. M., Knutton, S. and Frankel, G. (2005). Polarity of enteropathogenic *Escherichia coli* EspA filament assembly and protein secretion. *J. Bacteriol., 187,* 2881-2889.

[26] Buttner, D. and Bonas, U. (2002). Port of entry--the type III secretion translocon. *Trends Microbiol., 10,* 186-192.

[27] Blocker, A. J., Deane, J. E., Veenendaal, A. K., Roversi, P., Hodgkinson, J. L., Johnson, S., *et al.* (2008). What's the point of the type III secretion system needle? *Proc. Natl. Acad. Sci. USA, 105,* 6507-6513.

[28] Hayward, R. D., Cain, R. J., McGhie, E. J., Phillips, N., Garner, M. J. and Koronakis, V. (2005). Cholesterol binding by the bacterial type III translocon is essential for virulence effector delivery into mammalian cells. *Mol. Microbiol., 56,* 590-603.

[29] Faudry, E., Vernier, G., Neumann, E., Forge, V. and Attree, I. (2006). Synergistic pore formation by type III toxin translocators of *Pseudomonas aeruginosa*. *Biochemistry, 45,* 8117-8123.

[30] Viboud, G. I. and Bliska, J. B. (2001). A bacterial type III secretion system inhibits actin polymerization to prevent pore formation in host cell membranes. *EMBO J., 20,* 5373-5382.

[31] Aili, M., Isaksson, E. L., Carlsson, S. E., Wolf-Watz, H., Rosqvist, R. and Francis, M. S. (2008). Regulation of *Yersinia* Yop-effector delivery by translocated YopE. *Int. J. Med. Microbiol., 298,* 183-192.

[32] Cisz, M., Lee, P. C. and Rietsch, A. (2008). ExoS controls the cell contact-mediated switch to effector secretion in *Pseudomonas aeruginosa*. *J. Bacteriol., 190,* 2726-2738.

[33] Hartland, E. L. and Robins-Browne, R. M. (1998). In vitro association between the virulence proteins, YopD and YopE, of *Yersinia enterocolitica*. *FEMS Microbiol. Lett., 162,* 207-213.

[34] Abe, A., Matsuzawa, T. and Kuwae, A. (2005). Type-III effectors: sophisticated bacterial virulence factors. *C R Biol., 328,* 413-428.

[35] Galan, J. E. (2007). SnapShot: effector proteins of type III secretion systems. *Cell, 130,* 192.

[36] Lindeberg, M., Stavrinides, J., Chang, J. H., Alfano, J. R., Collmer, A., Dangl, J. L., *et al.* (2005). Proposed guidelines for a unified nomenclature and phylogenetic analysis of type III Hop effector proteins in the plant pathogen *Pseudomonas syringae*. *Mol Plant Microbe Interact, 18,* 275-282.

[37] McCann, H. C. and Guttman, D. S. (2008). Evolution of the type III secretion system and its effectors in plant-microbe interactions. *New Phytol., 177,* 33-47.

[38] Galan, J. E. (2009). Common themes in the design and function of bacterial effectors. *Cell Host Microbe., 5,* 571-579.

[39] Mattoo, S., Lee, Y. M. and Dixon, J. E. (2007). Interactions of bacterial effector proteins with host proteins. *Curr. Opin. Immunol., 19,* 392-401.

[40] Shames, S. R., Auweter, S. D. and Finlay, B. B. (2009). Co-evolution and exploitation of host cell signaling pathways by bacterial pathogens. *Int. J. Biochem. Cell Biol., 41,* 380-389.

[41] Stavrinides, J., McCann, H. C. and Guttman, D. S. (2008). Host-pathogen interplay and the evolution of bacterial effectors. *Cell. Microbiol., 10,* 285-292.

[42] Cunnac, S., Lindeberg, M. and Collmer, A. (2009). *Pseudomonas syringae* type III secretion system effectors: repertoires in search of functions. *Curr. Opin. Microbiol., 12,* 53-60.

[43] Block, A., Li, G., Fu, Z. Q. and Alfano, J. R. (2008). Phytopathogen type III effector weaponry and their plant targets. *Curr. Opin. Plant Biol., 11,* 396-403.

[44] Brutinel, E. D. and Yahr, T. L. (2008). Control of gene expression by type III secretory activity. *Curr. Opin. Microbiol., 11,* 128-133.

[45] Francis, M. S., Wolf-Watz, H. and Forsberg, Å. (2002). Regulation of type III secretion systems. *Curr. Opin. Microbiol., 5,* 166-172.

[46] Wickner, S., Maurizi, M. R. and Gottesman, S. (1999). Posttranslational quality control: folding, refolding, and degrading proteins. *Science, 286,* 1888-1893.

[47] Dougan, D. A., Mogk, A. and Bukau, B. (2002). Protein folding and degradation in bacteria: to degrade or not to degrade? That is the question. *Cell Mol. Life Sci., 59,* 1607-1616.

[48] Wick, L. M. and Egli, T. (2004). Molecular components of physiological stress responses in *Escherichia coli. Adv. Biochem. Eng. Biotechnol., 89,* 1-45.

[49] Guisbert, E., Yura, T., Rhodius, V. A. and Gross, C. A. (2008). Convergence of molecular, modeling, and systems approaches for an understanding of the *Escherichia coli* heat shock response. *Microbiol. Mol. Biol. Rev., 72,* 545-554.

[50] Gottesman, S. (1996). Proteases and their targets in *Escherichia coli. Annu. Rev. Genet, 30,* 465-506.

[51] Jackson, M. W., Silva-Herzog, E. and Plano, G. V. (2004). The ATP-dependent ClpXP and Lon proteases regulate expression of the *Yersinia*

pestis type III secretion system via regulated proteolysis of YmoA, a small histone-like protein. *Mol. Microbiol., 54,* 1364-1378.

[52] Bretz, J., Losada, L., Lisboa, K. and Hutcheson, S. W. (2002). Lon protease functions as a negative regulator of type III protein secretion in *Pseudomonas syringae*. *Mol. Microbiol., 45,* 397-409.

[53] Iyoda, S. and Watanabe, H. (2005). ClpXP protease controls expression of the type III protein secretion system through regulation of RpoS and GrlR levels in enterohemorrhagic *Escherichia coli*. *J. Bacteriol., 187,* 4086-4094.

[54] Boddicker, J. D. and Jones, B. D. (2004). Lon protease activity causes down-regulation of *Salmonella* pathogenicity island 1 invasion gene expression after infection of epithelial cells. *Infect. Immun., 72,* 2002-2013.

[55] Lan, L., Deng, X., Xiao, Y., Zhou, J. M. and Tang, X. (2007). Mutation of Lon protease differentially affects the expression of *Pseudomonas syringae* type III secretion system genes in rich and minimal media and reduces pathogenicity. *Mol. Plant. Microbe Interact, 20,* 682-696.

[56] Kage, H., Takaya, A., Ohya, M. and Yamamoto, T. (2008). Coordinated regulation of expression of *Salmonella* pathogenicity island 1 and flagellar type III secretion systems by ATP-dependent ClpXP protease. *J. Bacteriol., 190,* 2470-2478.

[57] Duguay, A. R. and Silhavy, T. J. (2004). Quality control in the bacterial periplasm. *Biochim. Biophys. Acta, 1694,* 121-134.

[58] Macritchie, D. M., Ward, J. D., Nevesinjac, A. Z. and Raivio, T. L. (2008). Activation of the Cpx envelope stress response down-regulates expression of several locus of enterocyte effacement-encoded genes in enteropathogenic *Escherichia coli*. *Infect. Immun., 76,* 1465-1475.

[59] Carlsson, K. E., Liu, J., Edqvist, P. J. and Francis, M. S. (2007). Extracytoplasmic-stress-responsive pathways modulate type III secretion in *Yersinia pseudotuberculosis*. *Infect Immun, 75,* 3913-3924.

[60] Humphreys, S., Rowley, G., Stevenson, A., Anjum, M. F., Woodward, M. J., Gilbert, S., *et al.* (2004). Role of the two-component regulator CpxAR in the virulence of *Salmonella enterica* serotype Typhimurium. *Infect. Immun., 72,* 4654-4661.

[61] Nakayama, S., Kushiro, A., Asahara, T., Tanaka, R., Hu, L., Kopecko, D. J., *et al.* (2003). Activation of *hilA* expression at low pH requires the signal sensor CpxA, but not the cognate response regulator CpxR, in *Salmonella enterica* serovar Typhimurium. *Microbiology, 149,* 2809-2817.

[62] Nakayama, S. and Watanabe, H. (1998). Identification of *cpxR* as a positive regulator essential for expression of the *Shigella sonnei virF* gene. *J. Bacteriol., 180,* 3522-3528.

[63] Nakayama, S. and Watanabe, H. (1995). Involvement of cpxA, a sensor of a two-component regulatory system, in the pH-dependent regulation of expression of *Shigella sonnei virF* gene. *J. Bacteriol., 177,* 5062-5069.

[64] Mitobe, J., Arakawa, E. and Watanabe, H. (2005). A sensor of the two-component system CpxA affects expression of the type III secretion system through posttranscriptional processing of InvE. *J. Bacteriol., 187,* 107-113.

[65] Osborne, S. E. and Coombes, B. K. (2009). RpoE fine tunes expression of a subset of SsrB-regulated virulence factors in *Salmonella enterica* serovar Typhimurium. *BMC Microbiol., 9,* 45.

[66] Ha, U. H., Wang, Y. and Jin, S. (2003). DsbA of *Pseudomonas aeruginosa* is essential for multiple virulence factors. *Infect Immun, 71,* 1590-1595.

[67] Jackson, M. W. and Plano, G. V. (1999). DsbA is required for stable expression of outer membrane protein YscC and for efficient Yop secretion in *Yersinia pestis. J. Bacteriol., 181,* 5126-5130.

[68] Miki, T., Okada, N. and Danbara, H. (2004). Two periplasmic disulfide oxidoreductases, DsbA and SrgA, target outer membrane protein SpiA, a component of the *Salmonella* pathogenicity island 2 type III secretion system. *J. Biol. Chem., 279,* 34631-34642.

[69] Watarai, M., Tobe, T., Yoshikawa, M. and Sasakawa, C. (1995). Disulfide oxidoreductase activity of *Shigella flexneri* is required for release of Ipa proteins and invasion of epithelial cells. *Proc. Natl. Acad. Sci. USA, 92,* 4927-4931.

[70] Yu, J., Edwards-Jones, B., Neyrolles, O. and Kroll, J. S. (2000). Key role for DsbA in cell-to-cell spread of *Shigella flexneri*, permitting secretion of Ipa proteins into interepithelial protrusions. *Infect. Immun., 68,* 6449-6456.

[71] Miki, T., Okada, N., Kim, Y., Abe, A. and Danbara, H. (2008). DsbA directs efficient expression of outer membrane secretin EscC of the enteropathogenic *Escherichia coli* type III secretion apparatus. *Microb. Pathog., 44,* 151-158.

[72] Augustin, D. K., Song, Y., Baek, M. S., Sawa, Y., Singh, G., Taylor, B., *et al.* (2007). Presence or absence of lipopolysaccharide O antigens affects type III secretion by *Pseudomonas aeruginosa. J. Bacteriol., 189,* 2203-2209.

[73] West, N. P., Sansonetti, P., Mounier, J., Exley, R. M., Parsot, C., Guadagnini, S., *et al.* (2005). Optimization of virulence functions through glucosylation of *Shigella* LPS. *Science, 307,* 1313-1317.

[74] Mota, L. J., Journet, L., Sorg, I., Agrain, C. and Cornelis, G. R. (2005). Bacterial injectisomes: needle length does matter. *Science, 307,* 1278.

[75] Fardini, Y., Trotereau, J., Bottreau, E., Souchard, C., Velge, P. and Virlogeux-Payant, I. (2009). Investigation of the role of the BAM complex and SurA chaperone in outer-membrane protein biogenesis and type III secretion system expression in *Salmonella. Microbiology, 155,* 1613-1622.

[76] Fardini, Y., Chettab, K., Grepinet, O., Rochereau, S., Trotereau, J., Harvey, P., *et al.* (2007). The YfgL lipoprotein is essential for type III secretion system expression and virulence of *Salmonella enterica* Serovar Enteritidis. *Infect. Immun., 75,* 358-370.

[77] Fadl, A. A., Galindo, C. L., Sha, J., Klimpel, G. R., Popov, V. L. and Chopra, A. K. (2006). Global gene expression of a murein (Braun) lipoprotein mutant of *Salmonella enterica* serovar Typhimurium by microarray analysis. *Gene, 374,* 121-127.

[78] Randall, L. L. and Hardy, S. J. (2002). SecB, one small chaperone in the complex milieu of the cell. *Cell Mol. Life Sci, 59,* 1617-1623.

[79] Zhou, J. and Xu, Z. (2005). The structural view of bacterial translocation-specific chaperone SecB: implications for function. *Mol. Microbiol., 58,* 349-357.

[80] Driessen, A. J. and Nouwen, N. (2008). Protein translocation across the bacterial cytoplasmic membrane. *Annu Rev Biochem, 77,* 643-667.

[81] Jones, S. E., Lloyd, L. J., Tan, K. K. and Buck, M. (2003). Secretion defects that activate the phage shock response of *Escherichia coli. J. Bacteriol., 185,* 6707-6711.

[82] Jovanovic, G., Lloyd, L. J., Stumpf, M. P., Mayhew, A. J. and Buck, M. (2006). Induction and function of the phage shock protein extracytoplasmic stress response in *Escherichia coli. J. Biol. Chem., 281,* 21147-21161.

[83] Kleerebezem, M., Crielaard, W. and Tommassen, J. (1996). Involvement of stress protein PspA (phage shock protein A) of *Escherichia coli* in maintenance of the protonmotive force under stress conditions. *EMBO J., 15,* 162-171.

[84] Darwin, A. J. (2007). Regulation of the phage-shock-protein stress response in *Yersinia enterocolitica. Adv .Exp. Med. Biol., 603,* 167-177.

[85] Wattiau, P. and Cornelis, G. R. (1993). SycE, a chaperone-like protein of *Yersinia enterocolitica* involved in the secretion of YopE. *Mol. Microbiol., 8,* 123-131.

[86] Feldman, M. F. and Cornelis, G. R. (2003). The multitalented type III chaperones: all you can do with 15 kDa. *FEMS Microbiol. Lett., 219,* 151-158.

[87] Page, A. L. and Parsot, C. (2002). Chaperones of the type III secretion pathway: jacks of all trades. *Mol Microbiol, 46,* 1-11.

[88] Parsot, C., Hamiaux, C. and Page, A. L. (2003). The various and varying roles of specific chaperones in type III secretion systems. *Curr. Opin. Microbiol, 6,* 7-14.

[89] Wattiau, P., Woestyn, S. and Cornelis, G. R. (1996). Customized secretion chaperones in pathogenic bacteria. *Mol. Microbiol., 20,* 255-262.

[90] Wilharm, G., Dittmann, S., Schmid, A. and Heesemann, J. (2007). On the role of specific chaperones, the specific ATPase, and the proton motive force in type III secretion. *Int. J. Med. Microbiol., 297,* 27-36.

[91] Van Eerde, A., Hamiaux, C., Perez, J., Parsot, C. and Dijkstra, B. W. (2004). Structure of Spa15, a type III secretion chaperone from *Shigella flexneri* with broad specificity. *EMBO Rep., 5,* 477-483.

[92] Buttner, C. R., Cornelis, G. R., Heinz, D. W. and Niemann, H. H. (2005). Crystal structure of *Yersinia enterocolitica* type III secretion chaperone SycT. *Protein. Sci., 14,* 1993-2002.

[93] Locher, M., Lehnert, B., Krauss, K., Heesemann, J., Groll, M. and Wilharm, G. (2005). Crystal structure of the *Yersinia enterocolitica* type III secretion chaperone SycT. *J. Biol. Chem, 280,* 31149-31155.

[94] Luo, Y., Bertero, M. G., Frey, E. A., Pfuetzner, R. A., Wenk, M. R., Creagh, L., *et al.* (2001). Structural and biochemical characterization of the type III secretion chaperones CesT and SigE. *Nat. Struct. Biol., 8,* 1031-1036.

[95] Trame, C. B. and McKay, D. B. (2003). Structure of the *Yersinia enterocolitica* molecular-chaperone protein SycE. *Acta Crystallogr D Biol Crystallogr, 59,* 389-392.

[96] Evdokimov, A. G., Tropea, J. E., Routzahn, K. M. and Waugh, D. S. (2002). Three-dimensional structure of the type III secretion chaperone SycE from *Yersinia pestis. Acta Crystallogr D Biol. Crystallogr, 58,* 398-406.

[97] Singer, A. U., Desveaux, D., Betts, L., Chang, J. H., Nimchuk, Z., Grant, S. R., *et al.* (2004). Crystal structures of the type III effector protein AvrPphF

and its chaperone reveal residues required for plant pathogenesis. *Structure, 12,* 1669-1681.

[98] Lilic, M., Vujanac, M. and Stebbins, C. E. (2006). A common structural motif in the binding of virulence factors to bacterial secretion chaperones. *Mol. Cell, 21,* 653-664.

[99] Stebbins, C. E. and Galán, J. E. (2001). Maintenance of an unfolded polypeptide by a cognate chaperone in bacterial type III secretion. *Nature, 414,* 77-81.

[100] Phan, J., Tropea, J. E. and Waugh, D. S. (2004). Structure of the *Yersinia pestis* type III secretion chaperone SycH in complex with a stable fragment of YscM2. *Acta Crystallogr D Biol. Crystallogr, 60,* 1591-1599.

[101] Birtalan, S. and Ghosh, P. (2001). Structure of the *Yersinia* type III secretory system chaperone SycE. *Nat. Struct .Biol., 8,* 974-978.

[102] Day, J. B. and Plano, G. V. (1998). A complex composed of SycN and YscB functions as a specific chaperone for YopN in *Yersinia pestis*. *Mol. Microbiol., 30,* 777-788.

[103] Schubot, F. D., Jackson, M. W., Penrose, K. J., Cherry, S., Tropea, J. E., Plano, G. V., *et al.* (2005). Three-dimensional structure of a macromolecular assembly that regulates type III secretion in *Yersinia pestis*. *J Mol Biol, 346,* 1147-1161.

[104] Pallen, M. J., Francis, M. S. and Futterer, K. (2003). Tetratricopeptide-like repeats in type-III-secretion chaperones and regulators. *FEMS Microbiol Lett, 223,* 53-60.

[105] Lamb, J. R., Tugendreich, S. and Hieter, P. (1995). Tetratrico peptide repeat interactions: to TPR or not to TPR? *Trends Biochem. Sci., 20,* 257-259.

[106] D'Andrea, L. D. and Regan, L. (2003). TPR proteins: the versatile helix. *Trends Biochem. Sci., 28,* 655-662.

[107] Blatch, G. L. and Lassle, M. (1999). The tetratricopeptide repeat: a structural motif mediating protein-protein interactions. *Bioessays, 21,* 932-939.

[108] Buttner, C. R., Sorg, I., Cornelis, G. R., Heinz, D. W. and Niemann, H. H. (2008). Structure of the *Yersinia enterocolitica* type III secretion translocator chaperone SycD. *J. Mol. Biol., 375,* 997-1012.

[109] Lunelli, M., Lokareddy, R. K., Zychlinsky, A. and Kolbe, M. (2009). IpaB-IpgC interaction defines binding motif for type III secretion translocator. *Proc. Natl. Acad. Sci .USA, 106,* 9661-9666.

[110] Tan, Y. W., Yu, H. B., Sivaraman, J., Leung, K. Y. and Mok, Y. K. (2009). Mapping of the chaperone AcrH binding regions of translocators AopB and

AopD and characterization of oligomeric and metastable AcrH-AopB-AopD complexes in the type III secretion system of *Aeromonas hydrophila. Protein Sci, 18,* 1724-1734.

[111] Schmid, A., Dittmann, S., Grimminger, V., Walter, S., Heesemann, J. and Wilharm, G. (2006). *Yersinia enterocolitica* type III secretion chaperone SycD: Recombinant expression, purification and characterization of a homodimer. *Protein Expr. Purif., 49,* 176-182.

[112] Darwin, K. H. and Miller, V. L. (2001). Type III secretion chaperone-dependent regulation: activation of virulence genes by SicA and InvF in *Salmonella typhimurium. EMBO J., 20,* 1850-1862.

[113] Tengel, T., Sethson, I. and Francis, M. S. (2002). Conformational analysis by CD and NMR spectroscopy of a peptide encompassing the amphipathic domain of YopD from *Yersinia. Eur. J. Biochem., 269,* 3659-3668.

[114] Creasey, E. A., Friedberg, D., Shaw, R. K., Umanski, T., Knutton, S., Rosenshine, I., *et al.* (2003). CesAB is an enteropathogenic *Escherichia coli* chaperone for the type-III translocator proteins EspA and EspB. *Microbiology, 149,* 3639-3647.

[115] Yip, C. K., Finlay, B. B. and Strynadka, N. C. (2005). Structural characterization of a type III secretion system filament protein in complex with its chaperone. *Nat. Struct. Mol. Biol, 12,* 75-81.

[116] Su, M. S., Kao, H. C., Lin, C. N. and Syu, W. J. (2008). Gene l0017 encodes a second chaperone for EspA of enterohaemorrhagic *Escherichia coli* O157 : H7. *Microbiology, 154,* 1094-1103.

[117] Neves, B. C., Mundy, R., Petrovska, L., Dougan, G., Knutton, S. and Frankel, G. (2003). CesD2 of enteropathogenic *Escherichia coli* is a second chaperone for the type III secretion translocator protein EspD. *Infect Immun, 71,* 2130-2141.

[118] Wainwright, L. A. and Kaper, J. B. (1998). EspB and EspD require a specific chaperone for proper secretion from enteropathogenic *Escherichia coli. Mol. Microbiol, 27,* 1247-1260.

[119] Daniell, S. J., Delahay, R. M., Shaw, R. K., Hartland, E. L., Pallen, M. J., Booy, F., *et al.* (2001). Coiled-coil domain of enteropathogenic *Escherichia coli* type III secreted protein EspD is involved in EspA filament-mediated cell attachment and hemolysis. *Infect. Immun, 69,* 4055-4064.

[120] Quinaud, M., Chabert, J., Faudry, E., Neumann, E., Lemaire, D., Pastor, A., *et al.* (2005). The PscE-PscF-PscG complex controls type III secretion needle biogenesis in *Pseudomonas aeruginosa. J. Biol. Chem, 280,* 36293-36300.

[121] Tan, Y. W., Yu, H. B., Leung, K. Y., Sivaraman, J. and Mok, Y. K. (2008). Structure of AscE and induced burial regions in AscE and AscG upon formation of the chaperone needle-subunit complex of type III secretion system in *Aeromonas hydrophila. Protein Sci, 17,* 1748-1760.

[122] Quinaud, M., Ple, S., Job, V., Contreras-Martel, C., Simorre, J. P., Attree, I., *et al.* (2007). Structure of the heterotrimeric complex that regulates type III secretion needle formation. *Proc. Natl. Acad. Sci. USA, 104,* 7803-7808.

[123] Sun, P., Tropea, J. E., Austin, B. P., Cherry, S. and Waugh, D. S. (2008). Structural characterization of the *Yersinia pestis* type III secretion system needle protein YscF in complex with its heterodimeric chaperone YscE/YscG. *J. Mol. Biol, 377,* 819-830.

[124] Lee, V. T., Tam, C. and Schneewind, O. (2000). LcrV, a substrate for *Yersinia enterocolitica* type III secretion, is required for toxin targeting into the cytosol of HeLa cells. *J. Biol. Chem, 275,* 36869-36875.

[125] Matson, J. S. and Nilles, M. L. (2001). LcrG-LcrV interaction is required for control of Yops secretion in *Yersinia pestis. J. Bacteriol, 183,* 5082-5091.

[126] Poueymiro, M. and Genin, S. (2009). Secreted proteins from *Ralstonia solanacearum*: a hundred tricks to kill a plant. *Curr. Opin. Microbiol, 12,* 44-52.

[127] Alfano, J. R. and Collmer, A. (2004). Type III secretion system effector proteins: double agents in bacterial disease and plant defense. *Annu. Rev. Phytopathol, 42,* 385-414.

[128] Lewis, J. D., Guttman, D. S. and Desveaux, D. (2009). The targeting of plant cellular systems by injected type III effector proteins. *Semin. Cell Dev. Biol*, doi:10.1016/j.semcdb.2009.06.003.

[129] Bonas, U. and Lahaye, T. (2002). Plant disease resistance triggered by pathogen-derived molecules: refined models of specific recognition. *Curr. Opin. Microbiol, 5,* 44-50.

[130] Grant, S. R., Fisher, E. J., Chang, J. H., Mole, B. M. and Dangl, J. L. (2006). Subterfuge and manipulation: type III effector proteins of phytopathogenic bacteria. *Annu. Re.v Microbiol., 60,* 425-449.

[131] Zhou, J. M. and Chai, J. (2008). Plant pathogenic bacterial type III effectors subdue host responses. *Curr. Opin. Microbiol, 11,* 179-185.

[132] McDowell, J. M. and Simon, S. A. (2008). Molecular diversity at the plant-pathogen interface. *Dev. Comp. Immunol., 32,* 736-744.

[133] Buttner, D. and Bonas, U. (2006). Who comes first? How plant pathogenic bacteria orchestrate type III secretion. *Curr. Opin. Microbiol., 9,* 193-200.

[134] Buttner, D., Gurlebeck, D., Noel, L. D. and Bonas, U. (2004). HpaB from *Xanthomonas campestris* pv. vesicatoria acts as an exit control protein in type III-dependent protein secretion. *Mol. Microbiol., 54,* 755-768.

[135] Buttner, D., Lorenz, C., Weber, E. and Bonas, U. (2006). Targeting of two effector protein classes to the type III secretion system by a HpaC- and HpaB-dependent protein complex from *Xanthomonas campestris* pv. vesicatoria. *Mol. Microbiol, 59,* 513-527.

[136] Schesser, K., Frithz-Lindsten, E. and Wolf-Watz, H. (1996). Delineation and mutational analysis of the *Yersinia pseudotuberculosis* YopE domains which mediate translocation across bacterial and eukaryotic cellular membranes. *J. Bacteriol., 178,* 7227-7233.

[137] Woestyn, S., Sory, M. P., Boland, A., Lequenne, O. and Cornelis, G. R. (1996). The cytosolic SycE and SycH chaperones of *Yersinia* protect the region of YopE and YopH involved in translocation across eukaryotic cell membranes. *Mol .Microbiol, 20,* 1261-1271.

[138] Thomas, N. A., Deng, W., Baker, N., Puente, J. and Finlay, B. B. (2007). Hierarchical delivery of an essential host colonization factor in enteropathogenic *Escherichia coli. J. Biol. Chem, 282,* 29634-29645.

[139] Birtalan, S. C., Phillips, R. M. and Ghosh, P. (2002). Three-dimensional secretion signals in chaperone-effector complexes of bacterial pathogens. *Mol. Cell, 9,* 971-980.

[140] Rodgers, L., Gamez, A., Riek, R. and Ghosh, P. (2008). The type III secretion chaperone SycE promotes a localized disorder-to-order transition in the natively unfolded effector YopE. *J. Biol. Chem, 283,* 20857-20863.

[141] Knodler, L. A., Bertero, M., Yip, C., Strynadka, N. C. and Steele-Mortimer, O. (2006). Structure-based mutagenesis of SigE verifies the importance of hydrophobic and electrostatic residues in type III chaperone function. *Mol. Microbiol, 62,* 928-940.

[142] Neyt, C. and Cornelis, G. R. (1999). Role of SycD, the chaperone of the *Yersinia* Yop translocators YopB and YopD. *Mol Microbiol, 31,* 143-156.

[143] Francis, M. S., Aili, M., Wiklund, M. L. and Wolf-Watz, H. (2000). A study of the YopD-LcrH interaction from *Yersinia pseudotuberculosis* reveals a role for hydrophobic residues within the amphipathic domain of YopD. *Mol. Microbiol., 38,* 85-102.

[144] Francis, M. S. and Wolf-Watz, H. (1998). YopD of *Yersinia pseudotuberculosis* is translocated into the cytosol of HeLa epithelial cells: evidence of a structural domain necessary for translocation. *Mol. Microbiol, 29,* 799-813.

[145] Harrington, A. T., Hearn, P. D., Picking, W. L., Barker, J. R., Wessel, A. and Picking, W. D. (2003). Structural characterization of the N terminus of IpaC from *Shigella flexneri*. *Infect. Immun, 71,* 1255-1264.

[146] Kim, B. H., Kim, H. G., Kim, J. S., Jang, J. I. and Park, Y. K. (2007). Analysis of functional domains present in the N-terminus of the SipB protein. *Microbiology, 153,* 2998-3008.

[147] Schoehn, G., Di Guilmi, A. M., Lemaire, D., Attree, I., Weissenhorn, W. and Dessen, A. (2003). Oligomerization of type III secretion proteins PopB and PopD precedes pore formation in *Pseudomonas*. *EMBO J., 22,* 4957-4967.

[148] Mounier, J., Popoff, M. R., Enninga, J., Frame, M. C., Sansonetti, P. J. and Van Nhieu, G. T. (2009). The IpaC carboxyterminal effector domain mediates Src-dependent actin polymerization during *Shigella* invasion of epithelial cells. *PLoS Pathog, 5,* e1000271.

[149] Terry, C. M., Picking, W. L., Birket, S. E., Flentie, K., Hoffman, B. M., Barker, J. R., *et al.* (2008). The C-terminus of IpaC is required for effector activities related to *Shigella* invasion of host cells. *Microb. Pathog, 45,* 282-289.

[150] Hayward, R. D. and Koronakis, V. (1999). Direct nucleation and bundling of actin by the SipC protein of invasive *Salmonella*. *EMBO J., 18,* 4926-4934.

[151] Chang, J., Myeni, S. K., Lin, T. L., Wu, C. C., Staiger, C. J. and Zhou, D. (2007). SipC multimerization promotes actin nucleation and contributes to *Salmonella*-induced inflammation. *Mol. Microbiol., 66,* 1548-1556.

[152] Chang, J., Chen, J. and Zhou, D. (2005). Delineation and characterization of the actin nucleation and effector translocation activities of *Salmonella* SipC. *Mol. Microbiol, 55,* 1379-1389.

[153] Zurawski, D. V. and Stein, M. A. (2004). The SPI2-encoded SseA chaperone has discrete domains required for SseB stabilization and export, and binds within the C-terminus of SseB and SseD. *Microbiology, 150,* 2055-2068.

[154] Bröms, J. E., Edqvist, P. J., Forsberg, Å. and Francis, M. S. (2006). Tetratricopeptide repeats are essential for PcrH chaperone function in *Pseudomonas aeruginosa* type III secretion. *FEMS Microbiol. Lett, 256,* 57-66.

[155] Edqvist, P. J., Bröms, J. E., Betts, H. J., Forsberg, Å., Pallen, M. J. and Francis, M. S. (2006). Tetratricopeptide repeats in the type-III-secretion

chaperone, LcrH: their role in substrate binding and secretion. *Mol. Microbiol, 59,* 31-44.

[156] Edqvist, P. J., Aili, M., Liu, J. and Francis, M. S. (2007). Minimal YopB and YopD translocator secretion by *Yersinia* is sufficient for Yop-effector delivery into target cells. *Microbes Infect, 9,* 224-233.

[157] Menard, R., Sansonetti, P., Parsot, C. and Vasselon, T. (1994). Extracellular association and cytoplasmic partitioning of the IpaB and IpaC invasins of *S. flexneri. Cell, 79,* 515-525.

[158] Bröms, J. E., Edqvist, P. J., Carlsson, K. E., Forsberg, Å. and Francis, M. S. (2005). Mapping of a YscY binding domain within the LcrH chaperone that is required for regulation of *Yersinia* type III secretion. *J. Bacteriol, 187,* 7738-7752.

[159] Swietnicki, W., O'Brien, S., Holman, K., Cherry, S., Brueggemann, E., Tropea, J. E., *et al.* (2004). Novel protein-protein interactions of the *Yersinia pestis* type III secretion system elucidated with a matrix analysis by surface plasmon resonance and mass spectrometry. *J. Biol. Chem, 279,* 38693-38700.

[160] Francis, M. S., Lloyd, S. A. and Wolf-Watz, H. (2001). The type III secretion chaperone LcrH co-operates with YopD to establish a negative, regulatory loop for control of Yop synthesis in *Yersinia pseudotuberculosis. Mol. Microbiol, 42,* 1075-1093.

[161] Evans, L. D. and Hughes, C. (2009). Selective binding of virulence type III export chaperones by FliJ escort orthologues InvI and YscO. *FEMS Microbiol. Lett, 293,* 292-297.

[162] Broms, J. E., Forslund, A. L., Forsberg, A. and Francis, M. S. (2003). PcrH of *Pseudomonas aeruginosa* is essential for secretion and assembly of the type III translocon. *J. Infect. Dis, 188,* 1909-1921.

[163] Tucker, S. C. and Galán, J. E. (2000). Complex function for SicA, a *Salmonella enterica* serovar typhimurium type III secretion-associated chaperone. *J. Bacteriol., 182,* 2262-2268.

[164] Ruiz-Albert, J., Mundy, R., Yu, X. J., Beuzon, C. R. and Holden, D. W. (2003). SseA is a chaperone for the SseB and SseD translocon components of the *Salmonella* pathogenicity-island-2-encoded type III secretion system. *Microbiology, 149,* 1103-1111.

[165] Zurawski, D. V. and Stein, M. A. (2003). SseA acts as the chaperone for the SseB component of the *Salmonella* Pathogenicity Island 2 translocon. *Mol. Microbiol, 47,* 1341-1351.

[166] Foultier, B., Troisfontaines, P., Vertommen, D., Marenne, M. N., Rider, M., Parsot, C., *et al.* (2003). Identification of substrates and chaperone from the *Yersinia enterocolitica* 1B Ysa type III secretion system. *Infect. Immun, 71,* 242-253.

[167] Frithz-Lindsten, E., Rosqvist, R., Johansson, L. and Forsberg, Å. (1995). The chaperone-like protein YerA of *Yersinia pseudotuberculosis* stabilizes YopE in the cytoplasm but is dispensible for targeting to the secretion loci. *Mol. Microbiol, 16,* 635-647.

[168] Page, A. L., Sansonetti, P. and Parsot, C. (2002). Spa15 of *Shigella flexneri*, a third type of chaperone in the type III secretion pathway. *Mol. Microbiol, 43,* 1533-1542.

[169] Fu, Y. and Galán, J. E. (1998). Identification of a specific chaperone for SptP, a substrate of the centisome 63 type III secretion system of *Salmonella typhimurium*. *J. Bacteriol, 180,* 3393-3399.

[170] Losada, L. C. and Hutcheson, S. W. (2005). Type III secretion chaperones of *Pseudomonas syringae* protect effectors from Lon-associated degradation. *Mol. Microbiol, 55,* 941-953.

[171] Lloyd, S. A., Norman, M., Rosqvist, R. and Wolf-Watz, H. (2001). *Yersinia* YopE is targeted for type III secretion by N-terminal, not mRNA, signals. *Mol. Microbiol, 39,* 520-531.

[172] Persson, C., Nordfelth, R., Holmström, A., Håkansson, S., Rosqvist, R. and Wolf-Watz, H. (1995). Cell-surface-bound *Yersinia* translocate the protein tyrosine phosphatase YopH by a polarized mechanism into the target cell. *Mol. Microbiol, 18,* 135-150.

[173] Letzelter, M., Sorg, I., Mota, L. J., Meyer, S., Stalder, J., Feldman, M., *et al.* (2006). The discovery of SycO highlights a new function for type III secretion effector chaperones. *EMBO J., 25,* 3223-3233.

[174] Roppenser, B., Roder, A., Hentschke, M., Ruckdeschel, K. and Aepfelbacher, M. (2009). *Yersinia enterocolitica* differentially modulates RhoG activity in host cells. *J. Cell. Sci, 122,* 696-705.

[175] Krall, R., Zhang, Y. and Barbieri, J. T. (2004). Intracellular membrane localization of *Pseudomonas* ExoS and *Yersinia* YopE in mammalian cells. *J. Biol. Chem, 279,* 2747-2753.

[176] Zhang, Y. and Barbieri, J. T. (2005). A leucine-rich motif targets *Pseudomonas aeruginosa* ExoS within mammalian cells. *Infect. Immun. 73,* 7938-7945.

[177] Cheng, L. W. and Schneewind, O. (1999). *Yersinia enterocolitica* type III secretion. On the role of SycE in targeting YopE into HeLa cells. *J. Biol. Chem, 274,* 22102-22108.

[178] Wang, Y., Ouellette, A. N., Egan, C. W., Rathinavelan, T., Im, W. and De Guzman, R. N. (2007). Differences in the electrostatic surfaces of the type III secretion needle proteins PrgI, BsaL, and MxiH. *J .Mol .Biol, 371,* 1304-1314.

[179] Deane, J. E., Roversi, P., Cordes, F. S., Johnson, S., Kenjale, R., Daniell, S., *et al.* (2006). Molecular model of a type III secretion system needle: Implications for host-cell sensing. *Proc. Natl. Acad. Sci. USA, 103,* 12529-12533.

[180] Zhang, L., Wang, Y., Picking, W. L., Picking, W. D. and De Guzman, R. N. (2006). Solution structure of monomeric BsaL, the type III secretion needle protein of *Burkholderia pseudomallei. J. Mol .Biol., 359,* 322-330.

[181] Anderson, D. M., Fouts, D. E., Collmer, A. and Schneewind, O. (1999). Reciprocal secretion of proteins by the bacterial type III machines of plant and animal pathogens suggests universal recognition of mRNA targeting signals. *Proc. Natl. Acad. Sci. USA, 96,* 12839-12843.

[182] Rossier, O., Wengelnik, K., Hahn, K. and Bonas, U. (1999). The *Xanthomonas* Hrp type III system secretes proteins from plant and mammalian bacterial pathogens. *Proc Natl Acad Sci USA, 96,* 9368-9373.

[183] Young, B. M. and Young, G. M. (2002). YplA is exported by the Ysc, Ysa, and flagellar type III secretion systems of *Yersinia enterocolitica. J. Bacteriol., 184,* 1324-1334.

[184] Young, G. M., Schmiel, D. H. and Miller, V. L. (1999). A new pathway for the secretion of virulence factors by bacteria: the flagellar export apparatus functions as a protein-secretion system. *Proc. Natl .Acad. Sci. USA, 96,* 6456-6461.

[185] Badea, L., Beatson, S. A., Kaparakis, M., Ferrero, R. L. and Hartland, E. L. (2009). Secretion of flagellin by the LEE-encoded type III secretion system of enteropathogenic *Escherichia coli. BMC Microbiol, 9,* 30.

[186] Matsumoto, H. and Young, G. M. (2009). Essential role of the SycP chaperone in type III secretion of the YspP effector. *J Bacteriol, 191,* 1703-1715.

[187] Miao, E. A. and Miller, S. I. (2000). A conserved amino acid sequence directing intracellular type III secretion by *Salmonella typhimurium. Proc. Natl. Acad. Sci. USA, 97,* 7539-7544.

[188] Patel, J. C. and Galan, J. E. (2005). Manipulation of the host actin cytoskeleton by *Salmonella*--all in the name of entry. *Curr. Opin. Microbiol, 8,* 10-15.

[189] Ramsden, A. E., Holden, D. W. and Mota, L. J. (2007). Membrane dynamics and spatial distribution of *Salmonella*-containing vacuoles. *Trends Microbiol, 15,* 516-524.

[190] Rhen, M. and Dorman, C. J. (2005). Hierarchical gene regulators adapt *Salmonella enterica* to its host milieus. *Int. J. Med. Microbiol, 294,* 487-502.

[191] Ehrbar, K., Winnen, B. and Hardt, W. D. (2006). The chaperone binding domain of SopE inhibits transport via flagellar and SPI-1 TTSS in the absence of InvB. *Mol. Microbiol, 59,* 248-264.

[192] Lee, S. H. and Galan, J. E. (2004). *Salmonella* type III secretion-associated chaperones confer secretion-pathway specificity. *Mol. Microbiol, 51,* 483-495.

[193] Wood, S. E., Jin, J. and Lloyd, S. A. (2008). YscP and YscU switch the substrate specificity of the *Yersinia* type III secretion system by regulating export of the inner rod protein YscI. *J .Bacteriol, 190,* 4252-4262.

[194] Agrain, C., Callebaut, I., Journet, L., Sorg, I., Paroz, C., Mota, L. J., *et al.* (2005). Characterization of a Type III secretion substrate specificity switch (T3S4) domain in YscP from *Yersinia enterocolitica. Mol. Microbiol, 56,* 54-67.

[195] Edqvist, P. J., Olsson, J., Lavander, M., Sundberg, L., Forsberg, Å., Wolf-Watz, H., *et al.* (2003). YscP and YscU regulate substrate specificity of the *Yersinia* type III secretion system. *J. Bacteriol, 185,* 2259-2266.

[196] Magdalena, J., Hachani, A., Chamekh, M., Jouihri, N., Gounon, P., Blocker, A., *et al.* (2002). Spa32 regulates a switch in substrate specificity of the type III secreton of *Shigella flexneri* from needle components to Ipa proteins. *J .Bacteriol, 184,* 3433-3441.

[197] Tamano, K., Katayama, E., Toyotome, T. and Sasakawa, C. (2002). *Shigella* Spa32 is an essential secretory protein for functional type III secretion machinery and uniformity of its needle length. *J. Bacteriol, 184,* 1244-1252.

[198] Marlovits, T. C., Kubori, T., Lara-Tejero, M., Thomas, D., Unger, V. M. and Galan, J. E. (2006). Assembly of the inner rod determines needle length in the type III secretion injectisome. *Nature, 441,* 637-640.

[199] Lorenz, C., Schulz, S., Wolsch, T., Rossier, O., Bonas, U. and Buttner, D. (2008). HpaC controls substrate specificity of the *Xanthomonas* type III secretion system. *PLoS Pathog, 4,* e1000094.

[200] Thomas, N. A. and Finlay, B. B. (2004). Pathogens: bacterial needles ruled to length and specificity. *Curr. Biol., 14,* R192-194.

[201] Minamino, T., Imada, K. and Namba, K. (2008). Mechanisms of type III protein export for bacterial flagellar assembly. *Mol Biosyst, 4,* 1105-1115.

[202] Waters, R. C., O'Toole, P. W. and Ryan, K. A. (2007). The FliK protein and flagellar hook-length control. *Protein Sci, 16,* 769-780.

[203] Cornelis, G. R., Agrain, C. and Sorg, I. (2006). Length control of extended protein structures in bacteria and bacteriophages. *Curr. Opin. Microbiol, 9,* 201-206.

[204] Wang, D., Roe, A. J., McAteer, S., Shipston, M. J. and Gally, D. L. (2008). Hierarchal type III secretion of translocators and effectors from *Escherichia coli* O157:H7 requires the carboxy terminus of SepL that binds to Tir. *Mol Microbiol, 69,* 1499-1512.

[205] Botteaux, A., Sory, M. P., Biskri, L., Parsot, C. and Allaoui, A. (2009). MxiC is secreted by and controls the substrate specificity of the *Shigella flexneri* type III secretion apparatus. *Mol. Microbiol, 71,* 449-460.

[206] Kubori, T. and Galán, J. E. (2002). *Salmonella* type III secretion-associated protein InvE controls translocation of effector proteins into host cells. *J. Bacteriol, 184,* 4699-4708.

[207] Yu, X. J., Liu, M. and Holden, D. W. (2004). SsaM and SpiC interact and regulate secretion of *Salmonella* pathogenicity island 2 type III secretion system effectors and translocators. *Mol. Microbiol, 54,* 604-619.

[208] Yu, X. J., Ruiz-Albert, J., Unsworth, K. E., Garvis, S., Liu, M. and Holden, D. W. (2002). SpiC is required for secretion of *Salmonella* Pathogenicity Island 2 type III secretion system proteins. *Cell Microbiol, 4,* 531-540.

[209] Deng, W., Li, Y., Hardwidge, P. R., Frey, E. A., Pfuetzner, R. A., Lee, S., *et al.* (2005). Regulation of type III secretion hierarchy of translocators and effectors in attaching and effacing bacterial pathogens. *Infect. Immun, 73,* 2135-2146.

[210] O'Connell, C. B., Creasey, E. A., Knutton, S., Elliott, S., Crowther, L. J., Luo, W., *et al.* (2004). SepL, a protein required for enteropathogenic *Escherichia coli* type III translocation, interacts with secretion component SepD. *Mol .Microbiol, 52,* 1613-1625.

[211] Fu, Z. Q., Guo, M. and Alfano, J. R. (2006). *Pseudomonas syringae* HrpJ is a type III secreted protein that is required for plant pathogenesis, injection of effectors, and secretion of the HrpZ1 Harpin. *J. Bacteriol, 188,* 6060-6069.

[212] Deng, W., Puente, J. L., Gruenheid, S., Li, Y., Vallance, B. A., Vazquez, A., *et al.* (2004). Dissecting virulence: systematic and functional analyses of a pathogenicity island. *Proc .Natl. Acad. Sci .USA, 101,* 3597-3602.

[213] Boyd, A. P., Lambermont, I. and Cornelis, G. R. (2000). Competition between the Yops of *Yersinia enterocolitica* for delivery into eukaryotic cells: role of the SycE chaperone binding domain of YopE. *J. Bacteriol., 182,* 4811-4821.

[214] Gauthier, A. and Finlay, B. B. (2003). Translocated intimin receptor and its chaperone interact with ATPase of the type III secretion apparatus of enteropathogenic *Escherichia coli. J. Bacteriol., 185,* 6747-6755.

[215] Thomas, N. A., Deng, W., Puente, J. L., Frey, E. A., Yip, C. K., Strynadka, N. C., *et al.* (2005). CesT is a multi-effector chaperone and recruitment factor required for the efficient type III secretion of both LEE- and non-LEE-encoded effectors of enteropathogenic *Escherichia coli. Mol. Microbiol, 57,* 1762-1779.

[216] Akeda, Y. and Galan, J. E. (2005). Chaperone release and unfolding of substrates in type III secretion. *Nature, 437,* 911-915.

[217] Mills, E., Baruch, K., Charpentier, X., Kobi, S. and Rosenshine, I. (2008). Real-time analysis of effector translocation by the type III secretion system of enteropathogenic *Escherichia coli. Cell Host Microbe, 3,* 104-113.

[218] Van Engelenburg, S. B. and Palmer, A. E. (2008). Quantification of real-time *Salmonella* effector type III secretion kinetics reveals differential secretion rates for SopE2 and SptP. *Chem. Biol, 15,* 619-628.

[219] Winnen, B., Schlumberger, M. C., Sturm, A., Schupbach, K., Siebenmann, S., Jenny, P., *et al.* (2008). Hierarchical effector protein transport by the *Salmonella Typhimurium* SPI-1 type III secretion system. *PLoS ONE, 3,* e2178.

[220] Enninga, J., Mounier, J., Sansonetti, P. and Tran Van Nhieu, G. (2005). Secretion of type III effectors into host cells in real time. *Nat Methods, 2,* 959-965.

[221] Miller, V. L. (2002). Connections between transcriptional regulation and type III secretion? *Curr Opin Microbiol, 5,* 211-215.

[222] McCaw, M. L., Lykken, G. L., Singh, P. K. and Yahr, T. L. (2002). ExsD is a negative regulator of the *Pseudomonas aeruginosa* type III secretion regulon. *Mol. Microbiol, 46,* 1123-1133.

[223] Urbanowski, M. L., Lykken, G. L. and Yahr, T. L. (2005). A secreted regulatory protein couples transcription to the secretory activity of the

Pseudomonas aeruginosa type III secretion system. *Proc. Natl. Acad. Sci. USA, 102,* 9930-9935.

[224] Rietsch, A., Vallet-Gely, I., Dove, S. L. and Mekalanos, J. J. (2005). ExsE, a secreted regulator of type III secretion genes in *Pseudomonas aeruginosa. Proc. Natl. Acad. Sci .USA, 102,* 8006-8011.

[225] Dasgupta, N., Lykken, G. L., Wolfgang, M. C. and Yahr, T. L. (2004). A novel anti-anti-activator mechanism regulates expression of the *Pseudomonas aeruginosa* type III secretion system. *Mol. Microbiol, 53,* 297-308.

[226] Urbanowski, M. L., Brutinel, E. D. and Yahr, T. L. (2007). Translocation of ExsE into Chinese hamster ovary cells is required for transcriptional induction of the *Pseudomonas aeruginosa* type III secretion system. *Infect. Immun, 75,* 4432-4439.

[227] Brutinel, E. D., Vakulskas, C. A., Brady, K. M. and Yahr, T. L. (2008). Characterization of ExsA and of ExsA-dependent promoters required for expression of the *Pseudomonas aeruginosa* type III secretion system. *Mol. Microbiol, 68,* 657-671.

[228] Darwin, K. H. and Miller, V. L. (2000). The putative invasion protein chaperone SicA acts together with InvF to activate the expression of *Salmonella typhimurium* virulence genes. *Mol. Microbiol, 35,* 949-960.

[229] Mavris, M., Page, A. L., Tournebize, R., Demers, B., Sansonetti, P. and Parsot, C. (2002). Regulation of transcription by the activity of the *Shigella flexneri* type III secretion apparatus. *Mol. Microbiol, 43,* 1543-1553.

[230] Mavris, M., Sansonetti, P. J. and Parsot, C. (2002). Identification of the cis-acting site involved in activation of promoters regulated by activity of the type III secretion apparatus in *Shigella flexneri. J. Bacteriol, 184,* 6751-6759.

[231] Pilonieta, M. C. and Munson, G. P. (2008). The chaperone IpgC copurifies with the virulence regulator MxiE. *J Bacteriol, 190,* 2249-2251.

[232] Walker, K. A. and Miller, V. L. (2004). Regulation of the Ysa type III secretion system of *Yersinia enterocolitica* by YsaE/SycB and YsrS/YsrR. *J. Bacteriol, 186,* 4056-4066.

[233] Parsot, C., Ageron, E., Penno, C., Mavris, M., Jamoussi, K., d'Hauteville, H., *et al.* (2005). A secreted anti-activator, OspD1, and its chaperone, Spa15, are involved in the control of transcription by the type III secretion apparatus activity in *Shigella flexneri. Mol. Microbiol, 56,* 1627-1635.

[234] Stainier, I., Iriarte, M. and Cornelis, G. R. (1997). YscM1 and YscM2, two *Yersinia enterocolitica* proteins causing downregulation of *yop* transcription. *Mol. Microbiol, 26,* 833-843.

[235] Rimpiläinen, M., Forsberg, Å. and Wolf-Watz, H. (1992). A novel protein, LcrQ, involved in the low-calcium response of *Yersinia pseudotuberculosis* shows extensive homology to YopH. *J. Bacteriol, 174,* 3355-3363.

[236] Wulff-Strobel, C. R., Williams, A. W. and Straley, S. C. (2002). LcrQ and SycH function together at the Ysc type III secretion system in *Yersinia pestis* to impose a hierarchy of secretion. *Mol. Microbiol, 43,* 411-423.

[237] Cambronne, E. D. and Schneewind, O. (2002). *Yersinia enterocolitica* type III secretion: *yscM1* and *yscM2* regulate *yop* gene expression by a posttranscriptional mechanism that targets the 5' untranslated region of *yop* mRNA. *J. Bacteriol, 184,* 5880-5893.

[238] Cambronne, E. D., Cheng, L. W. and Schneewind, O. (2000). LcrQ/YscM1, regulators of the Yersinia *yop* virulon, are injected into host cells by a chaperone-dependent mechanism. *Mol. Microbiol, 37,* 263-273.

[239] Cambronne, E. D., Sorg, J. A. and Schneewind, O. (2004). Binding of SycH chaperone to YscM1 and YscM2 activates effector *yop* expression in *Yersinia enterocolitica. J. Bacteriol, 186,* 829-841.

[240] Anderson, D. M., Ramamurthi, K. S., Tam, C. and Schneewind, O. (2002). YopD and LcrH regulate expression of *Yersinia enterocolitica* YopQ by a posttranscriptional mechanism and bind to *yopQ* RNA. *J. Bacteriol, 184,* 1287-1295.

[241] Williams, A. W. and Straley, S. C. (1998). YopD of *Yersinia pestis* plays a role in negative regulation of the low-calcium response in addition to its role in translocation of Yops. *J. Bacteriol, 180,* 350-358.

[242] Dittmann, S., Schmid, A., Richter, S., Trulzsch, K., Heesemann, J. and Wilharm, G. (2007). The *Yersinia enterocolitica* type three secretion chaperone SycO is integrated into the Yop regulatory network and binds to the Yop secretion protein YscM1. *BMC Microbiol, 7,* 67.

[243] Schmid, A., Neumayer, W., Trulzsch, K., Israel, L., Imhof, A., Roessle, M., *et al.* (2009). Cross-talk between type three secretion system and metabolism in *Yersinia. J. Biol. Chem, 284,* 12165-12177.

[244] Dacheux, D., Epaulard, O., de Groot, A., Guery, B., Leberre, R., Attree, I., *et al.* (2002). Activation of the *Pseudomonas aeruginosa* type III secretion system requires an intact pyruvate dehydrogenase *aceAB* operon. *Infect. Immun, 70,* 3973-3977.

[245] Rietsch, A., Wolfgang, M. C. and Mekalanos, J. J. (2004). Effect of metabolic imbalance on expression of type III secretion genes in *Pseudomonas aeruginosa. Infect. Immun, 72,* 1383-1390.

[246] Wang, Q., Frye, J. G., McClelland, M. and Harshey, R. M. (2004). Gene expression patterns during swarming in *Salmonella typhimurium*: genes specific to surface growth and putative new motility and pathogenicity genes. *Mol. Microbiol., 52,* 169-187.

[247] Kim, W. and Surette, M. G. (2004). Metabolic differentiation in actively swarming *Salmonella. Mol Microbiol, 54,* 702-714.

[248] Rietsch, A. and Mekalanos, J. J. (2006). Metabolic regulation of type III secretion gene expression in *Pseudomonas aeruginosa. Mol. Microbiol, 59,* 807-820.

[249] Wolfgang, M. C., Lee, V. T., Gilmore, M. E. and Lory, S. (2003). Coordinate regulation of bacterial virulence genes by a novel adenylate cyclase-dependent signaling pathway. *Dev. Cell, 4,* 253-263.

[250] Petersen, S. and Young, G. M. (2002). Essential role for cyclic AMP and its receptor protein in *Yersinia enterocolitica* virulence. *Infect. Immun, 70,* 3665-3672.

[251] Marlovits, T. C., Kubori, T., Sukhan, A., Thomas, D. R., Galan, J. E. and Unger, V. M. (2004). Structural insights into the assembly of the type III secretion needle complex. *Science, 306,* 1040-1042.

[252] Galan, J. E. (2008). Energizing type III secretion machines: what is the fuel? *Nat Struct. Mol. Biol, 15,* 127-128.

[253] Sory, M. P., Boland, A., Lambermont, I. and Cornelis, G. R. (1995). Identification of the YopE and YopH domains required for secretion and internalization into the cytosol of macrophages, using the *cyaA* gene fusion approach. *Proc. Natl. Acad. Sci. USA, 92,* 11998-12002.

[254] Sory, M. P. and Cornelis, G. R. (1994). Translocation of a hybrid YopE-adenylate cyclase from *Yersinia enterocolitica* into HeLa cells. *Mol. Microbiol, 14,* 583-594.

[255] Lloyd, S. A., Sjöström, M., Andersson, S. and Wolf-Watz, H. (2002). Molecular characterization of type III secretion signals via analysis of synthetic N-terminal amino acid sequences. *Mol. Microbiol, 43,* 51-59.

[256] Anderson, D. M. and Schneewind, O. (1997). A mRNA signal for the type III secretion of Yop proteins by *Yersinia enterocolitica. Science, 278,* 1140-1143.

[257] Anderson, D. M. and Schneewind, O. (1999). *Yersinia enterocolitica* type III secretion: an mRNA signal that couples translation and secretion of YopQ. *Mol. Microbiol., 31,* 1139-1148.

[258] Russmann, H., Kubori, T., Sauer, J. and Galán, J. E. (2002). Molecular and functional analysis of the type III secretion signal of the *Salmonella enterica* InvJ protein. *Mol .Microbiol, 46,* 769-779.

[259] Broms, J. E., Francis, M. S. and Forsberg, A. (2007). Diminished LcrV secretion attenuates *Yersinia pseudotuberculosis* virulence. *J. Bacteriol, 189,* 8417-8429.

[260] Ramamurthi, K. S. and Schneewind, O. (2003). Substrate recognition by the *Yersinia* type III protein secretion machinery. *Mol. Microbiol, 50,* 1095-1102.

[261] Sorg, J. A., Miller, N. C. and Schneewind, O. (2005). Substrate recognition of type III secretion machines--testing the RNA signal hypothesis. *Cell Microbiol, 7,* 1217-1225.

[262] Lloyd, S. A., Forsberg, Å., Wolf-Watz, H. and Francis, M. S. (2001). Targeting exported substrates to the *Yersinia* TTSS: different functions for different signals? *Trends Microbiol, 9,* 367-371.

[263] Cornelis, G. R. (2003). How Yops find their way out of *Yersinia. Mol. Microbiol, 50,* 1091-1094.

[264] Cheng, L. W., Anderson, D. M. and Schneewind, O. (1997). Two independent type III secretion mechanisms for YopE in *Yersinia enterocolitica. Mol. Microbiol, 24,* 757-765.

[265] Sorg, J. A., Blaylock, B. and Schneewind, O. (2006). Secretion signal recognition by YscN, the *Yersinia* type III secretion ATPase. *Proc. Natl. Acad. Sci. USA, 103,* 16490-16495.

[266] Eichelberg, K., Ginocchio, C. C. and Galan, J. E. (1994). Molecular and functional characterization of the *Salmonella typhimurium* invasion genes *invB* and *invC*: homology of InvC to the F0F1 ATPase family of proteins. *J. Bacteriol, 176,* 4501-4510.

[267] Woestyn, S., Allaoui, A., Wattiau, P. and Cornelis, G. R. (1994). YscN, the putative energizer of the *Yersinia* Yop secretion machinery. *J. Bacteriol, 176,* 1561-1569.

[268] Blaylock, B., Riordan, K. E., Missiakas, D. M. and Schneewind, O. (2006). Characterization of the *Yersinia enterocolitica* type III secretion ATPase YscN and its regulator, YscL. *J .Bacteriol, 188,* 3525-3534.

[269] Pozidis, C., Chalkiadaki, A., Gomez-Serrano, A., Stahlberg, H., Brown, I., Tampakaki, A. P., *et al.* (2003). Type III protein translocase: HrcN is a

peripheral ATPase that is activated by oligomerization. *J. Biol. Chem, 278,* 25816-25824.

[270] Andrade, A., Pardo, J. P., Espinosa, N., Perez-Hernandez, G. and Gonzalez-Pedrajo, B. (2007). Enzymatic characterization of the enteropathogenic *Escherichia coli* type III secretion ATPase EscN. *Arch. Biochem. Biophys, 468,* 121-127.

[271] Lorenz, C. and Buttner, D. (2009). Functional characterization of the type III secretion ATPase HrcN from the plant pathogen *Xanthomonas campestris* pv. vesicatoria. *J. Bacteriol, 191,* 1414-1428.

[272] Muller, S. A., Pozidis, C., Stone, R., Meesters, C., Chami, M., Engel, A., *et al.* (2006). Double hexameric ring assembly of the type III protein translocase ATPase HrcN. *Mol Microbiol, 61,* 119-125.

[273] Zarivach, R., Vuckovic, M., Deng, W., Finlay, B. B. and Strynadka, N. C. (2007). Structural analysis of a prototypical ATPase from the type III secretion system. *Nat. Struct. Mol. Biol, 14,* 131-137.

[274] Jouihri, N., Sory, M. P., Page, A. L., Gounon, P., Parsot, C. and Allaoui, A. (2003). MxiK and MxiN interact with the Spa47 ATPase and are required for transit of the needle components MxiH and MxiI, but not of Ipa proteins, through the type III secretion apparatus of *Shigella flexneri. Mol. Microbiol, 49,* 755-767.

[275] Akeda, Y. and Galan, J. E. (2004). Genetic analysis of the *Salmonella enterica* type III secretion-associated ATPase InvC defines discrete functional domains. *J. Bacteriol, 186,* 2402-2412.

[276] Moraes, T. F., Spreter, T. and Strynadka, N. C. (2008). Piecing together the type III injectisome of bacterial pathogens. *Curr. Opin. Struct. Biol., 18,* 258-266.

[277] Lee, V. T. and Schneewind, O. (2002). Yop fusions to tightly folded protein domains and their effects on *Yersinia enterocolitica* type III secretion. *J. Bacteriol, 184,* 3740-3745.

[278] Feldman, M. F., Muller, S., Wuest, E. and Cornelis, G. R. (2002). SycE allows secretion of YopE-DHFR hybrids by the *Yersinia enterocolitica* type III Ysc system. *Mol. Microbiol, 46,* 1183-1197.

[279] Wilharm, G., Lehmann, V., Neumayer, W., Trcek, J. and Heesemann, J. (2004). *Yersinia enterocolitica* type III secretion: evidence for the ability to transport proteins that are folded prior to secretion. *BMC Microbiol, 4,* 27.

[280] Wilharm, G., Lehmann, V., Krauss, K., Lehnert, B., Richter, S., Ruckdeschel, K., *et al.* (2004). *Yersinia enterocolitica* type III secretion

depends on the proton motive force but not on the flagellar motor components MotA and MotB. *Infect. Immun, 72,* 4004-4009.

[281] Stebbins, C. E. and Galan, J. E. (2003). Priming virulence factors for delivery into the host. *Nat. Rev. Mol. Cell. Biol, 4,* 738-743.

[282] Chevance, F. F. and Hughes, K. T. (2008). Coordinating assembly of a bacterial macromolecular machine. *Nat. Rev. Microbiol, 6,* 455-465.

[283] Macnab, R. M. (2003). How bacteria assemble flagella. *Annu. Rev. Microbiol, 57,* 77-100.

[284] Kubori, T., Matsushima, Y., Nakamura, D., Uralil, J., Lara-Tejero, M., Sukhan, A., *et al.* (1998). Supramolecular structure of the *Salmonella typhimurium* type III protein secretion system. *Science, 280,* 602-605.

[285] Tamano, K., Aizawa, S., Katayama, E., Nonaka, T., Imajoh-Ohmi, S., Kuwae, A., *et al.* (2000). Supramolecular structure of the *Shigella* type III secretion machinery: the needle part is changeable in length and essential for delivery of effectors. *EMBO J., 19,* 3876-3887.

[286] Sekiya, K., Ohishi, M., Ogino, T., Tamano, K., Sasakawa, C. and Abe, A. (2001). Supermolecular structure of the enteropathogenic *Escherichia coli* type III secretion system and its direct interaction with the EspA-sheath-like structure. *Proc. Natl. Acad. Sci. USA, 98,* 11638-11643.

[287] Blocker, A., Gounon, P., Larquet, E., Niebuhr, K., Cabiaux, V., Parsot, C., *et al.* (1999). The tripartite type III secreton of *Shigella flexneri* inserts IpaB and IpaC into host membranes. *J. Cell Biol, 147,* 683-693.

[288] Blocker, A., Jouihri, N., Larquet, E., Gounon, P., Ebel, F., Parsot, C., *et al.* (2001). Structure and composition of the *Shigella flexneri* "needle complex", a part of its type III secreton. *Mol .Microbiol, 39,* 652-663.

[289] Sani, M., Allaoui, A., Fusetti, F., Oostergetel, G. T., Keegstra, W. and Boekema, E. J. (2007). Structural organization of the needle complex of the type III secretion apparatus of *Shigella flexneri. Micron, 38,* 291-301.

[290] Saier, M. H., Jr. (2004). Evolution of bacterial type III protein secretion systems. *Trends Microbiol, 12,* 113-115.

[291] Auvray, F., Thomas, J., Fraser, G. M. and Hughes, C. (2001). Flagellin polymerisation control by a cytosolic export chaperone. *J. Mol. Biol, 308,* 221-229.

[292] Muskotal, A., Kiraly, R., Sebestyen, A., Gugolya, Z., Vegh, B. M. and Vonderviszt, F. (2006). Interaction of FliS flagellar chaperone with flagellin. *FEBS Lett, 580,* 3916-3920.

[293] Ozin, A. J., Claret, L., Auvray, F. and Hughes, C. (2003). The FliS chaperone selectively binds the disordered flagellin C-terminal D0 domain central to polymerisation. *FEMS Microbiol. Lett, 219,* 219-224.

[294] Bennett, J. C., Thomas, J., Fraser, G. M. and Hughes, C. (2001). Substrate complexes and domain organization of the *Salmonella* flagellar export chaperones FlgN and FliT. *Mol. Microbiol, 39,* 781-791.

[295] Fraser, G. M., Bennett, J. C. and Hughes, C. (1999). Substrate-specific binding of hook-associated proteins by FlgN and FliT, putative chaperones for flagellum assembly. *Mol. Microbiol, 32,* 569-580.

[296] Evdokimov, A. G., Phan, J., Tropea, J. E., Routzahn, K. M., Peters, H. K., Pokross, M., *et al.* (2003). Similar modes of polypeptide recognition by export chaperones in flagellar biosynthesis and type III secretion. *Nat. Struct. Biol, 10,* 789-793.

[297] Evans, L. D., Stafford, G. P., Ahmed, S., Fraser, G. M. and Hughes, C. (2006). An escort mechanism for cycling of export chaperones during flagellum assembly. *Proc. Natl. Acad. Sci. USA, 103,* 17474-17479.

[298] Thomas, J., Stafford, G. P. and Hughes, C. (2004). Docking of cytosolic chaperone-substrate complexes at the membrane ATPase during flagellar type III protein export. *Proc. Natl. Acad .Sci. USA, 101,* 3945-3950.

[299] Soutourina, O. A. and Bertin, P. N. (2003). Regulation cascade of flagellar expression in Gram-negative bacteria. *FEMS Microbiol Rev, 27,* 505-523.

[300] McCarter, L. L. (2006). Regulation of flagella. *Curr. Opin. Microbiol, 9,* 180-186.

[301] Chilcott, G. S. and Hughes, K. T. (2000). Coupling of flagellar gene expression to flagellar assembly in *Salmonella enterica* serovar typhimurium and *Escherichia coli. Microbiol. Mol. Biol. Rev., 64,* 694-708.

[302] Aldridge, P. D., Karlinsey, J. E., Aldridge, C., Birchall, C., Thompson, D., Yagasaki, J., *et al.* (2006). The flagellar-specific transcription factor, sigma28, is the Type III secretion chaperone for the flagellar-specific anti-sigma28 factor FlgM. *Genes .Dev, 20,* 2315-2326.

[303] Yamamoto, S. and Kutsukake, K. (2006). FliT acts as an anti-FlhD2C2 factor in the transcriptional control of the flagellar regulon in *Salmonella enterica* serovar typhimurium. *J. Bacteriol, 188,* 6703-6708.

[304] Aldridge, P., Karlinsey, J. and Hughes, K. T. (2003). The type III secretion chaperone FlgN regulates flagellar assembly via a negative feedback loop containing its chaperone substrates FlgK and FlgL. *Mol. Microbiol, 49,* 1333-1345.

[305] Karlinsey, J. E., Lonner, J., Brown, K. L. and Hughes, K. T. (2000). Translation/secretion coupling by type III secretion systems. *Cell, 102,* 487-497.

[306] Hughes, K. T., Gillen, K. L., Semon, M. J. and Karlinsey, J. E. (1993). Sensing structural intermediates in bacterial flagellar assembly by export of a negative regulator. *Science, 262,* 1277-1280.

[307] Juhas, M., Crook, D. W. and Hood, D. W. (2008). Type IV secretion systems: tools of bacterial horizontal gene transfer and virulence. *Cell Microbiol, 10,* 2377-2386.

[308] Backert, S. and Meyer, T. F. (2006). Type IV secretion systems and their effectors in bacterial pathogenesis. *Curr Opin Microbiol, 9,* 207-217.

[309] Baron, C. (2005). From bioremediation to biowarfare: on the impact and mechanism of type IV secretion systems. *FEMS Microbiol. Lett, 253,* 163-170.

[310] Christie, P. J., Atmakuri, K., Krishnamoorthy, V., Jakubowski, S. and Cascales, E. (2005). Biogenesis, architecture, and function of bacterial type IV secretion systems. *Annu. Rev. Microbiol, 59,* 451-485.

[311] Cambronne, E. D. and Roy, C. R. (2006). Recognition and delivery of effector proteins into eukaryotic cells by bacterial secretion systems. *Traffic, 7,* 929-939.

[312] Gomis-Ruth, F. X., Sola, M., de la Cruz, F. and Coll, M. (2004). Coupling factors in macromolecular type-IV secretion machineries. *Curr. Pharm. Des, 10,* 1551-1565.

[313] Deng, W., Chen, L., Peng, W. T., Liang, X., Sekiguchi, S., Gordon, M. P., *et al.* (1999). VirE1 is a specific molecular chaperone for the exported single-stranded-DNA-binding protein VirE2 in *Agrobacterium*. *Mol. Microbiol, 31,* 1795-1807.

[314] Sundberg, C., Meek, L., Carroll, K., Das, A. and Ream, W. (1996). VirE1 protein mediates export of the single-stranded DNA-binding protein VirE2 from *Agrobacterium tumefaciens* into plant cells. *J. Bacteriol., 178,* 1207-1212.

[315] Zhao, Z., Sagulenko, E., Ding, Z. and Christie, P. J. (2001). Activities of *virE1* and the VirE1 secretion chaperone in export of the multifunctional VirE2 effector via an *Agrobacterium* type IV secretion pathway. *J. Bacteriol., 183,* 3855-3865.

[316] Sundberg, C. D. and Ream, W. (1999). The *Agrobacterium tumefaciens* chaperone-like protein, VirE1, interacts with VirE2 at domains required for

single-stranded DNA binding and cooperative interaction. *J. Bacteriol., 181,* 6850-6855.

[317] Duckely, M., Oomen, C., Axthelm, F., Van Gelder, P., Waksman, G. and Engel, A. (2005). The VirE1VirE2 complex of *Agrobacterium tumefaciens* interacts with single-stranded DNA and forms channels. *Mol. Microbiol., 58,* 1130-1142.

[318] Vergunst, A. C., van Lier, M. C., den Dulk-Ras, A. and Hooykaas, P. J. (2003). Recognition of the *Agrobacterium tumefaciens* VirE2 translocation signal by the VirB/D4 transport system does not require VirE1. *Plant Physiol, 133,* 978-988.

[319] Atmakuri, K., Ding, Z. and Christie, P. J. (2003). VirE2, a type IV secretion substrate, interacts with the VirD4 transfer protein at cell poles of *Agrobacterium tumefaciens. Mol. Microbiol., 49,* 1699-1713.

[320] Dym, O., Albeck, S., Unger, T., Jacobovitch, J., Branzburg, A., Michael, Y.*, et al.* (2008). Crystal structure of the *Agrobacterium* virulence complex VirE1-VirE2 reveals a flexible protein that can accommodate different partners. *Proc. Natl. Acad. Sci. USA, 105,* 11170-11175.

[321] Cambronne, E. D. and Roy, C. R. (2007). The *Legionella pneumophila* IcmSW complex interacts with multiple Dot/Icm effectors to facilitate type IV translocation. *PLoS Pathog, 3,* e188.

[322] Al-Khodor, S., Price, C. T., Habyarimana, F., Kalia, A. and Abu Kwaik, Y. (2008). A Dot/Icm-translocated ankyrin protein of *Legionella pneumophila* is required for intracellular proliferation within human macrophages and protozoa. *Mol. Microbiol, 70,* 908-923.

[323] Bardill, J. P., Miller, J. L. and Vogel, J. P. (2005). IcmS-dependent translocation of SdeA into macrophages by the *Legionella pneumophila* type IV secretion system. *Mol. Microbiol, 56,* 90-103.

[324] Ninio, S., Celli, J. and Roy, C. R. (2009). A *Legionella pneumophila* effector protein encoded in a region of genomic plasticity binds to Dot/Icm-modified vacuoles. *PLoS Pathog, 5,* e1000278.

[325] Ninio, S., Zuckman-Cholon, D. M., Cambronne, E. D. and Roy, C. R. (2005). The *Legionella* IcmS-IcmW protein complex is important for Dot/Icm-mediated protein translocation. *Mol. Microbiol, 55,* 912-926.

[326] Vincent, C. D. and Vogel, J. P. (2006). The *Legionella pneumophila* IcmS-LvgA protein complex is important for Dot/Icm-dependent intracellular growth. *Mol. Microbiol, 61,* 596-613.

[327] Coers, J., Kagan, J. C., Matthews, M., Nagai, H., Zuckman, D. M. and Roy, C. R. (2000). Identification of Icm protein complexes that play distinct roles

in the biogenesis of an organelle permissive for *Legionella pneumophila* intracellular growth. *Mol. Microbiol, 38,* 719-736.

[328] Dumenil, G. and Isberg, R. R. (2001). The *Legionella pneumophila* IcmR protein exhibits chaperone activity for IcmQ by preventing its participation in high-molecular-weight complexes. *Mol .Microbiol, 40,* 1113-1127.

[329] Raychaudhury, S., Farelli, J. D., Montminy, T. P., Matthews, M., Menetret, J. F., Dumenil, G., *et al.* (2009). Structure and function of interacting IcmR-IcmQ domains from a type IVb secretion system in *Legionella pneumophila. Structure, 17,* 590-601.

[330] Dumenil, G., Montminy, T. P., Tang, M. and Isberg, R. R. (2004). IcmR-regulated membrane insertion and efflux by the *Legionella pneumophila* IcmQ protein. *J .Biol .Chem, 279,* 4686-4695.

[331] Dumas, F., Duckely, M., Pelczar, P., Van Gelder, P. and Hohn, B. (2001). An *Agrobacterium* VirE2 channel for transferred-DNA transport into plant cells. *Proc. Natl. Acad. Sci. USA, 98,* 485-490.

[332] Pattis, I., Weiss, E., Laugks, R., Haas, R. and Fischer, W. (2007). The *Helicobacter pylori* CagF protein is a type IV secretion chaperone-like molecule that binds close to the C-terminal secretion signal of the CagA effector protein. *Microbiology, 153,* 2896-2909.

[333] Couturier, M. R., Tasca, E., Montecucco, C. and Stein, M. (2006). Interaction with CagF is required for translocation of CagA into the host via the *Helicobacter pylori* type IV secretion system. *Infect. Immun., 74,* 273-281.

[334] Filloux, A., Hachani, A. and Bleves, S. (2008). The bacterial type VI secretion machine: yet another player for protein transport across membranes. *Microbiology, 154,* 1570-1583.

[335] Cascales, E. (2008). The type VI secretion toolkit. *EMBO Rep, 9,* 735-741.

[336] Pukatzki, S., McAuley, S. B. and Miyata, S. T. (2009). The type VI secretion system: translocation of effectors and effector-domains. *Curr. Opin. Microbiol, 12,* 11-17.

[337] Shrivastava, S. and Mande, S. S. (2008). Identification and functional characterization of gene components of Type VI Secretion system in bacterial genomes. *PLoS ONE, 3,* e2955.

[338] de Bruin, O. M., Ludu, J. S. and Nano, F. E. (2007). The *Francisella* pathogenicity island protein IglA localizes to the bacterial cytoplasm and is needed for intracellular growth. *BMC Microbiol, 7,* 1.

[339] Zheng, J. and Leung, K. Y. (2007). Dissection of a type VI secretion system in *Edwardsiella tarda. Mol .Microbiol., 66,* 1192-1206.

[340] Schlieker, C., Zentgraf, H., Dersch, P. and Mogk, A. (2005). ClpV, a unique Hsp100/Clp member of pathogenic proteobacteria. *Biol. Chem., 386,* 1115-1127.

[341] Filloux, A. (2009). The type VI secretion system: a tubular story. *EMBO J., 28,* 309-310.

[342] Bonemann, G., Pietrosiuk, A., Diemand, A., Zentgraf, H. and Mogk, A. (2009). Remodelling of VipA/VipB tubules by ClpV-mediated threading is crucial for type VI protein secretion. *EMBO J., 28,* 315-325.

[343] Ellis, R. J. (1997). Do molecular chaperones have to be proteins? *Biochem Biophys. Res. Commun, 238,* 687-692.

[344] Henderson, B., Allan, E. and Coates, A. R. (2006). Stress wars: the direct role of host and bacterial molecular chaperones in bacterial infection. *Infect. Immun, 74,* 3693-3706.

[345] Neckers, L. and Tatu, U. (2008). Molecular chaperones in pathogen virulence: emerging new targets for therapy. *Cell Host Microbe, 4,* 519-527.

[346] Aberg, V. and Almqvist, F. (2007). Pilicides-small molecules targeting bacterial virulence. *Org. Biomol. Chem., 5,* 1827-1834.

[347] Keyser, P., Elofsson, M., Rosell, S. and Wolf-Watz, H. (2008). Virulence blockers as alternatives to antibiotics: type III secretion inhibitors against Gram-negative bacteria. *J. Intern. Med, 264,* 17-29.

[348] Clatworthy, A. E., Pierson, E. and Hung, D. T. (2007). Targeting virulence: a new paradigm for antimicrobial therapy. *Nat. Chem. Biol., 3,* 541-548.

[349] Wattiau, P., Bernier, B., Deslee, P., Michiels, T. and Cornelis, G. R. (1994). Individual chaperones required for Yop secretion by *Yersinia. Proc. Natl. Acad. Sci. USA, 91,* 10493-10497.

[350] Iriarte, M. and Cornelis, G. R. (1998). YopT, a new *Yersinia* Yop effector protein, affects the cytoskeleton of host cells. *Mol Microbiol, 29,* 915-929.

[351] Cheng, L. W., Kay, O. and Schneewind, O. (2001). Regulated secretion of YopN by the type III machinery of *Yersinia enterocolitica. J. Bacteriol., 183,* 5293-5301.

[352] Ferracci, F., Schubot, F. D., Waugh, D. S. and Plano, G. V. (2005). Selection and characterization of *Yersinia pestis* YopN mutants that constitutively block Yop secretion. *Mol. Microbiol., 57,* 970-987.

[353] Jackson, M. W., Day, J. B. and Plano, G. V. (1998). YscB of *Yersinia pestis* functions as a specific chaperone for YopN. *J Bacteriol, 180,* 4912-4921.

[354] Day, J. B., Guller, I. and Plano, G. V. (2000). *Yersinia pestis* YscG protein is a Syc-like chaperone that directly binds YscE. *Infect. Immun., 68,* 6466-6471.

[355] Phan, J., Austin, B. P. and Waugh, D. S. (2005). Crystal structure of the *Yersinia* type III secretion protein YscE. *Protein Sci., 14,* 2759-2763.

[356] Day, J. B. and Plano, G. V. (2000). The *Yersinia pestis* YscY protein directly binds YscX, a secreted component of the type III secretion machinery. *J. Bacteriol., 182,* 1834-1843.

[357] Iriarte, M. and Cornelis, G. R. (1999). Identification of SycN, YscX, and YscY, three new elements of the *Yersinia yop* virulon. *J. Bacteriol, 181,* 675-680.

[358] Shen, D. K., Quenee, L., Bonnet, M., Kuhn, L., Derouazi, M., Lamotte, D., *et al.* (2008). Orf1/SpcS chaperones ExoS for type three secretion by *Pseudomonas aeruginosa. Biomed Environ. Sci., 21,* 103-109.

[359] Finck-Barbancon, V., Yahr, T. L. and Frank, D. W. (1998). Identification and characterization of SpcU, a chaperone required for efficient secretion of the ExoU cytotoxin. *J. Bacteriol., 180,* 6224-6231.

[360] Allmond, L. R., Karaca, T. J., Nguyen, V. N., Nguyen, T., Wiener-Kronish, J. P. and Sawa, T. (2003). Protein binding between PcrG-PcrV and PcrH-PopB/PopD encoded by the *pcrGVH-popBD* operon of the *Pseudomonas aeruginosa* type III secretion system. *Infect. Immun., 71,* 2230-2233.

[361] Faudry, E., Job, V., Dessen, A., Attree, I. and Forge, V. (2007). Type III secretion system translocator has a molten globule conformation both in its free and chaperone-bound forms. *FEBS J., 274,* 3601-3610.

[362] Yang, H., Shan, Z., Kim, J., Wu, W., Lian, W., Zeng, L., *et al.* (2007). Regulatory role of PopN and its interacting partners in type III secretion of *Pseudomonas aeruginosa. J. Bacteriol., 189,* 2599-2609.

[363] Niebuhr, K., Jouihri, N., Allaoui, A., Gounon, P., Sansonetti, P. J. and Parsot, C. (2000). IpgD, a protein secreted by the type III secretion machinery of *Shigella flexneri,* is chaperoned by IpgE and implicated in entry focus formation. *Mol. Microbiol, 38,* 8-19.

[364] Ogawa, M., Suzuki, T., Tatsuno, I., Abe, H. and Sasakawa, C. (2003). IcsB, secreted via the type III secretion system, is chaperoned by IpgA and required at the post-invasion stage of *Shigella* pathogenicity. *Mol. Microbiol, 48,* 913-931.

[365] Hachani, A., Biskri, L., Rossi, G., Marty, A., Menard, R., Sansonetti, P., *et al.* (2008). IpgB1 and IpgB2, two homologous effectors secreted via the Mxi-Spa type III secretion apparatus, cooperate to mediate polarized cell invasion and inflammatory potential of *Shigella flexenri. Microbes Infect, 10,* 260-268.

[366] Birket, S. E., Harrington, A. T., Espina, M., Smith, N. D., Terry, C. M., Darboe, N., *et al.* (2007). Preparation and characterization of translocator/chaperone complexes and their component proteins from *Shigella flexneri*. *Biochemistry, 46,* 8128-8137.

[367] Page, A. L., Ohayon, H., Sansonetti, P. J. and Parsot, C. (1999). The secreted IpaB and IpaC invasins and their cytoplasmic chaperone IpgC are required for intercellular dissemination of *Shigella flexneri*. *Cell Microbiol, 1,* 183-193.

[368] Darwin, K. H., Robinson, L. S. and Miller, V. L. (2001). SigE is a chaperone for the *Salmonella enterica* serovar Typhimurium invasion protein SigD. *J. Bacteriol, 183,* 1452-1454.

[369] Bronstein, P. A., Miao, E. A. and Miller, S. I. (2000). InvB is a type III secretion chaperone specific for SspA. *J. Bacteriol, 182,* 6638-6644.

[370] Ehrbar, K., Friebel, A., Miller, S. I. and Hardt, W. D. (2003). Role of the *Salmonella* pathogenicity island 1 (SPI-1) protein InvB in type III secretion of SopE and SopE2, two *Salmonella* effector proteins encoded outside of SPI-1. *J .Bacteriol, 185,* 6950-6967.

[371] Ehrbar, K., Hapfelmeier, S., Stecher, B. and Hardt, W. D. (2004). InvB is required for type III-dependent secretion of SopA in *Salmonella enterica* serovar Typhimurium. *J. Bacteriol, 186,* 1215-1219.

[372] Lee, S. H. and Galan, J. E. (2003). InvB is a type III secretion-associated chaperone for the *Salmonella enterica* effector protein SopE. *J. Bacteriol, 185,* 7279-7284.

[373] Higashide, W. and Zhou, D. (2006). The first 45 amino acids of SopA are necessary for InvB binding and SPI-1 secretion. *J Bacteriol, 188,* 2411-2420.

[374] Dai, S. and Zhou, D. (2004). Secretion and function of *Salmonella* SPI-2 effector SseF require its chaperone, SscB. *J. Bacteriol, 186,* 5078-5086.

[375] Elliott, S. J., O'Connell, C. B., Koutsouris, A., Brinkley, C., Donnenberg, M. S., Hecht, G., *et al.* (2002). A gene from the locus of enterocyte effacement that is required for enteropathogenic *Escherichia coli* to increase tight-junction permeability encodes a chaperone for EspF. *Infect. Immun, 70,* 2271-2277.

[376] Abe, A., de Grado, M., Pfuetzner, R. A., Sanchez-Sanmartin, C., Devinney, R., Puente, J. L., *et al.* (1999). Enteropathogenic *Escherichia coli* translocated intimin receptor, Tir, requires a specific chaperone for stable secretion. *Mol. Microbiol, 33,* 1162-1175.

[377] Creasey, E. A., Delahay, R. M., Bishop, A. A., Shaw, R. K., Kenny, B., Knutton, S., *et al.* (2003). CesT is a bivalent enteropathogenic *Escherichia coli* chaperone required for translocation of both Tir and Map. *Mol .Microbiol, 47,* 209-221.

[378] Elliott, S. J., Hutcheson, S. W., Dubois, M. S., Mellies, J. L., Wainwright, L. A., Batchelor, M., *et al.* (1999). Identification of CesT, a chaperone for the type III secretion of Tir in enteropathogenic *Escherichia coli. Mol. Microbiol, 33,* 1176-1189.

[379] Li, M., Rosenshine, I., Yu, H. B., Nadler, C., Mills, E., Hew, C. L., *et al.* (2006). Identification and characterization of NleI, a new non-LEE-encoded effector of enteropathogenic *Escherichia coli* (EPEC). *Microbes Infect, 8,* 2890-2898.

[380] Betts, H. J., Twiggs, L. E., Sal, M. S., Wyrick, P. B. and Fields, K. A. (2008). Bioinformatic and biochemical evidence for the identification of the type III secretion system needle protein of *Chlamydia trachomatis. J. Bacteriol, 190,* 1680-1690.

[381] Fields, K. A., Fischer, E. R., Mead, D. J. and Hackstadt, T. (2005). Analysis of putative *Chlamydia trachomatis* chaperones Scc2 and Scc3 and their use in the identification of type III secretion substrates. *J. Bacteriol, 187,* 6466-6478.

[382] Slepenkin, A., de la Maza, L. M. and Peterson, E. M. (2005). Interaction between components of the type III secretion system of Chlamydiaceae. *J. Bacteriol, 187,* 473-479.

[383] Panina, E. M., Mattoo, S., Griffith, N., Kozak, N. A., Yuk, M. H. and Miller, J. F. (2005). A genome-wide screen identifies a *Bordetella* type III secretion effector and candidate effectors in other species. *Mol. Microbiol, 58,* 267-279.

[384] Zheng, J., Li, N., Tan, Y. P., Sivaraman, J., Mok, Y. K., Mo, Z. L., *et al.* (2007). EscC is a chaperone for the *Edwardsiella tarda* type III secretion system putative translocon components EseB and EseD. *Microbiology, 153,* 1953-1962.

[385] Wang, B., Mo, Z. L., Mao, Y. X., Zou, Y. X., Xiao, P., Li, J., *et al.* (2009). Investigation of EscA as a chaperone for the *Edwardsiella tarda* type III secretion system putative translocon component EseC. *Microbiology, 155,* 1260-1271.

[386] Akeda, Y., Okayama, K., Kimura, T., Dryselius, R., Kodama, T., Oishi, K., *et al.* (2009). Identification and characterization of a type III secretion-

associated chaperone in the type III secretion system 1 of *Vibrio parahaemolyticus*. *FEMS Microbiol. Lett, 296,* 18-25.

[387] Gaudriault, S., Paulin, J. P. and Barny, M. A. (2002). The DspB/F protein of *Erwinia amylovora* is a type III secretion chaperone ensuring efficient intrabacterial production of the Hrp-secreted DspA/E pathogenicity factor. *Mol. Plant Pathol, 3,* 313-320.

[388] Gaudriault, S., Malandrin, L., Paulin, J. P. and Barny, M. A. (1997). DspA, an essential pathogenicity factor of *Erwinia amylovora* showing homology with AvrE of *Pseudomonas syringae*, is secreted via the Hrp secretion pathway in a DspB-dependent way. *Mol. Microbiol, 26,* 1057-1069.

[389] van Dijk, K., Tam, V. C., Records, A. R., Petnicki-Ocwieja, T. and Alfano, J. R. (2002). The ShcA protein is a molecular chaperone that assists in the secretion of the HopPsyA effector from the type III (Hrp) protein secretion system of *Pseudomonas syringae*. *Mol. Microbiol, 44,* 1469-1481.

[390] Charity, J. C., Pak, K., Delwiche, C. F. and Hutcheson, S. W. (2003). Novel exchangeable effector loci associated with the *Pseudomonas syringae hrp* pathogenicity island: evidence for integron-like assembly from transposed gene cassettes. *Mol .Plant Microbe Interact, 16,* 495-507.

[391] Bogdanove, A. J., Kim, J. F., Wei, Z., Kolchinsky, P., Charkowski, A. O., Conlin, A. K., *et al.* (1998). Homology and functional similarity of an *hrp*-linked pathogenicity locus, *dspEF*, of *Erwinia amylovora* and the avirulence locus *avrE* of *Pseudomonas syringae* pathovar tomato. *Proc. Natl. Acad. Sci. USA, 95,* 1325-1330.

[392] Shan, L., Oh, H. S., Chen, J., Guo, M., Zhou, J., Alfano, J. R., *et al.* (2004). The HopPtoF locus of *Pseudomonas syringae* pv. tomato DC3000 encodes a type III chaperone and a cognate effector. *Mol. Plant. Microbe Interact, 17,* 447-455.

[393] Wehling, M. D., Guo, M., Fu, Z. Q. and Alfano, J. R. (2004). The *Pseudomonas syringae* HopPtoV protein is secreted in culture and translocated into plant cells via the type III protein secretion system in a manner dependent on the ShcV type III chaperone. *J. Bacteriol., 186,* 3621-3630.

[394] Badel, J. L., Nomura, K., Bandyopadhyay, S., Shimizu, R., Collmer, A. and He, S. Y. (2003). *Pseudomonas syringae* pv. tomato DC3000 HopPtoM (CEL ORF3) is important for lesion formation but not growth in tomato and is secreted and translocated by the Hrp type III secretion system in a chaperone-dependent manner. *Mol. Microbiol., 49,* 1239-1251.

[395] Kabisch, U., Landgraf, A., Krause, J., Bonas, U. and Boch, J. (2005). Type III secretion chaperones ShcS1 and ShcO1 from *Pseudomonas syringae* pv. tomato DC3000 bind more than one effector. *Microbiology, 151,* 269-280.

[396] Guo, M., Chancey, S. T., Tian, F., Ge, Z., Jamir, Y. and Alfano, J. R. (2005). *Pseudomonas syringae* type III chaperones ShcO1, ShcS1, and ShcS2 facilitate translocation of their cognate effectors and can substitute for each other in the secretion of HopO1-1. *J. Bacteriol, 187,* 4257-4269.

[397] Wei, C. F., Deng, W. L. and Huang, H. C. (2005). A chaperone-like HrpG protein acts as a suppressor of HrpV in regulation of the *Pseudomonas syringae* pv. syringae type III secretion system. *Mol. Microbiol, 57,* 520-536.

[398] Lorenz, C., Kirchner, O., Egler, M., Stuttmann, J., Bonas, U. and Buttner, D. (2008). HpaA from *Xanthomonas* is a regulator of type III secretion. *Mol. Microbiol, 69,* 344-360.

[399] He, S. Y., Nomura, K. and Whittam, T. S. (2004). Type III protein secretion mechanism in mammalian and plant pathogens. *Biochim. Biophys. Acta, 1694,* 181-206.

[400] Kubori, T., Sukhan, A., Aizawa, S. I. and Galán, J. E. (2000). Molecular characterization and assembly of the needle complex of the *Salmonella typhimurium* type III protein secretion system. *Proc. Natl. Acad. Sci. USA, 97,* 10225-10230.

[401] Shaw, R. K., Daniell, S., Ebel, F., Frankel, G. and Knutton, S. (2001). EspA filament-mediated protein translocation into red blood cells. *Cell Microbiol, 3,* 213-222.

[402] Guex, N. and Peitsch, M. C. (1997). SWISS-MODEL and the Swiss-PdbViewer: an environment for comparative protein modeling. *Electrophoresis, 18,* 2714-2723.

[403] Potterton, L., McNicholas, S., Krissinel, E., Gruber, J., Cowtan, K., Emsley, P., *et al.* (2004). Developments in the CCP4 molecular-graphics project. *Acta Crystallogr D Biol. Crystallogr, 60,* 2288-2294.

[404] Aldridge, P. and Hughes, K. T. (2001). How and when are substrates selected for type III secretion? *Trends Microbiol, 9,* 209-214.

Index

D

N

O

P